YACHT CRUISING

Other Books by Patrick Ellam

SOPRANINO

THE SPORTSMAN'S GUIDE
TO THE CARIBBEAN

UNDERWATER

SAM & ME

WIND SONG

YACHT CRUISING

Patrick Ellam

W·W·NORTON & COMPANY
NEW YORK LONDON

Published simultaneously in Canada by George J. McLeod Limited, Toronto.
Printed in the United States of America.

The text of this book is composed in Times Roman, with display type set in Baskerville. Composition by ComCom. Manufacturing by Haddon Craftsmen. Book design by M. Franklyn Plympton.

Library of Congress Cataloging in Publication Data
Ellam, Patrick.
Yacht cruising.
Includes index.
1. Sailing. 2. Seamanship. 3. Yachts and yachting. I. Title.
GV811.5.E54 1983 797.1'24 82-22472

ISBN 0-393-03280-9

W. W. Norton & Company, Inc., 500 Fifth Avenue, New York, N. Y. 10110
W. W. Norton & Company Ltd., 37 Great Russell Street, London WC1B 3NU

For those who want to
take their yachts farther
than they have before

Contents

Illustrations

Preface

In writing this book we tried to include all the details that, taken together, give an old hand a bag of tricks that he can dip into when conventional procedures are inadequate. Life being what it is, these various tricks would not all fit into neat chapters. But we have put the miscellaneous ones in wherever they would go and hope that the reader will find them, with the help of the index.

Patrick Ellam

Ventura, California
January 1983

1

Checking the Compass

The most important instrument aboard a yacht is her main compass. Not only is it used to steer her but it also serves as the standard against which her other compasses are verified.

So even if it was swung recently, we always check the main compass ourselves when first we take a boat out, in case there has been any change in the magnetic field around it. To check the compass we must find a place on the chart where there are suitable ranges.

A range may be a lighthouse with a headland behind it, or a church spire in line with a factory chimney, or any two marks like that. You can even use a buoy as the mark nearest to you, if the other one is so far away that a slight variation in the buoy's position will not matter. Each range can be used to take readings over the bow and over the stern of the boat. But there must be enough of them to check the compass in every direction.

If possible, the place you choose should also be in calm, slack water, for that makes the job much easier. When the ranges have been selected, draw them on the chart, with the magnetic bearing (in both directions) beside each one. Then go there in the boat and take your readings.

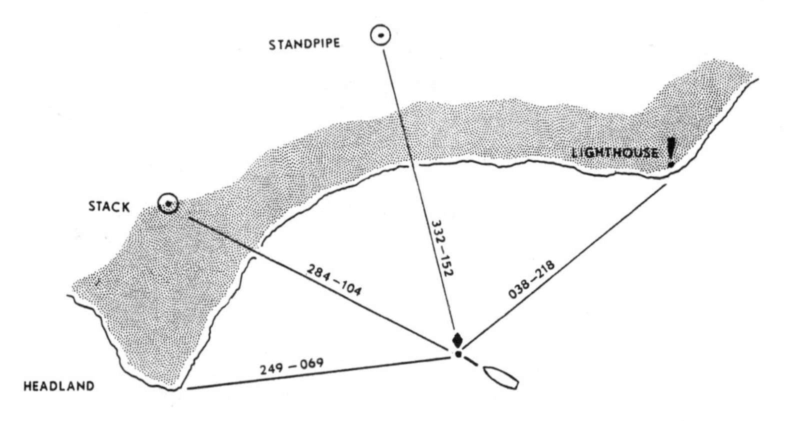

Checking the compass: The chart has been prepared by drawing the ranges on it and marking their bearings each way. Now the boat is lined up with the buoy and the stack for the first shot.

The idea is to bring the two marks of a range into line, then aim the bow at them and read the compass. But in real life, it is not that simple. For when the marks come into line, her bow is usually pointing somewhere else. And by the time you get her aimed at the range, the marks have separated, as she drifts sideways on the wind or current. So you go a few yards the other way, then aim her bow at the farthest mark and keep it there as she drifts back. Meanwhile someone else watches the compass and, as the marks come into line, takes the reading. This method works well, even in quite difficult conditions.

When you have taken all your readings over the bow, you should be ready to tackle the ones over the stern. For those you face aft, while the man reading the compass faces forward. But there are two problems: It is harder to aim over the stern, because there is less boat to aim with. And when you turn the wheel to your right, the stern goes to your left. But after a few tries, you get the hang of it. If a reading looks doubtful, you can always do it again.

When you have all the readings, make your own correction card. Use a circular diagram, as illustrated in the drawing, instead of a conventional deviation table. The advantage of the circular layout is that it does not require readings at fixed intervals. Any reading you can get, you can use. And we find a correction card (which tells you what corrections to apply) more useful than a deviation table (which says what the errors are).

Making the card is easy. When you compare a reading you have just taken with the magnetic bearing of that range, the correction you will have to apply in the future is obvious. Mark that on the card, opposite the heading. When you use the card, it is not hard to interpolate for any other heading between the ones for which you have corrections.

So if your card were like the drawing and you wanted to proceed on a heading of 057 degrees (magnetic), you would apply a correction of + 10° and steer 067 degrees. Or if you wanted to head due west, you would use a correction of – 6° and steer 264 degrees. And so on, for any other heading.

But whenever we get the chance, we keep checking to make sure that nothing has changed and that the card is still accurate. For example, the navigational aids leading out of a port will often give you a new correction to add to those on your card. As long as this correction fits neatly

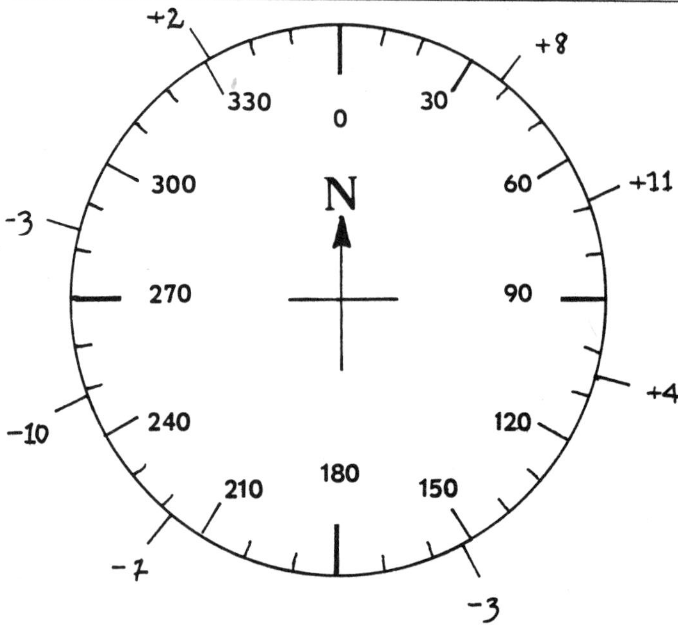

A compass correction card of the kind that we use. Inside the circle are the headings and outside are the corrections, opposite the headings to which they refer. Corrections for intermediate headings are found by interpolation.

between the others, you have no problem. But if not, you should be highly suspicious.

Has someone moved an anchor from one cockpit locker to another recently? Or stowed a flashlight near the compass? Or put an iron key in the binnacle? Whatever has happened, you must find it and correct it, then check the compass again to see that it has reverted to its old form.

This extra work is worth the effort to be sure of your compass. If you cannot be certain of your compass, you can never trust your navigation.

2

Radio Direction Finding

Often you hear people say that radio direction finders are not to be trusted, yet we have found them to be among the most reliable navigational instruments aboard yachts. Many times we have made coastwise passages of several days' duration—in fog all the way—where a direction finder has been our primary source of information. How did we do it? By installing the instrument carefully in the boat, learning what it can and can not do, then operating it patiently and meticulously within those limits.

There are many kinds of radio direction finder but the most common type is a squarish box with a ferrite antenna that you can turn on top of it, so we'll start with that.

Rule Number One—in our book—is NEVER TAKE IT ON DECK. The instruments don't like the damp and the operator will end up peering uselessly into the fog, when he should be attending to business.

So first of all, install the radio direction finder down below, in a permanent location, if possible, even if it has to be upside down on the overhead. (You soon learn to work it that way up.) Otherwise, it has to have a "nest" where it lives and a set of chocks in which you place it when you are about to use it. But the chocks must be so tight that it can not twist more than a degree or so, at the most, or you can forget the whole thing. For at a range of sixty miles, which is normal, a degree makes a difference of slightly over a mile. So a five-degree error in the location of the instrument will put you five miles out of position and prevent you from ever getting a decent fix.

When installing a direction finder, it is best to line it up parallel with the centerline of the boat—as closely as you can—since that will reduce the correction you have to make later.

Basically the instrument is nothing but a radio receiver with a movable directional antenna to find where the signal is coming from. With that type of movable antenna the signal strength is almost constant except for two places—180 degrees apart—where there are dead spots, called *nulls.* So you use those nulls to find the bearing of the transmitting station. In ideal conditions a null will be so narrow and so sharply defined that you can easily tell exactly where it is. But more often it is a vague area, several degrees wide, in which the signal dies down and comes back up again.

The volume control should work smoothly without sudden dips in the sound level and with no rattling noises as you turn it. Some sets also have a gain control that lets you vary the signal amplification to get a clearer null.

The switch position marked *BFO* turns on a beat frequency oscillator that will convert a dull, hissing signal into a clear beep. It should be left off when not needed, since it tends to distort an otherwise good signal. And that is about the extent of the gadgets we find useful.

The null meter that is on some sets is all very well in ideal conditions, when there is only one signal to be heard and that is loud and clear. But more often you hear a confused mess of signals, sounding like ten birds in a distant tree. And the one you want, of course, is chirping away quietly between two louder ones. So we never use the meter, except to check the battery.

Neither do we use the sensing devices that are often provided to tell you which side of a station you are on, since in most cases that is obvious and on the rare occasions when it is not, a second bearing, taken a few minutes later, will clear the matter up. The navigator's life is complicated enough without adding any gadgets he doesn't need.

If the set does not check out, first try a new battery. If that does not cure it, take it to a reliable service man who has the equipment to test it properly. Like any radio, it could have a component that has failed. Sometimes we have found a broken ferrite rod in the antenna.

Finally, start the boat's engine to see if it interferes with the direction finder. If it does, you have the choice of fitting surpressors to the engine or turning the engine off when you use the set.

Now you can take the boat out to sea and swing her, to determine the errors in the system. By "the system," we mean the radio direction finder plus the boat's main compass. We will use them both together to find the bearing of the transmitting station. So you need two people: one to operate the direction finder, the other to steer the boat and read the main compass.

Radio signals often change direction slightly when they cross a coastline at an angle, so the station used should have a clear shot to the place you select for swinging. Since the transmitter sends out signals in all directions, you may well pick up echoes that have bounced off nearby objects, such as cliffs or ships, if you are too close to the station. So we never take radio bearings from less than six miles away. And lastly—in choosing a place to check the system, select a location where you are sure of your position, such as near a small navigation buoy that won't upset anything.

Now the idea is to take a series of bearings on the station, with the boat on different headings, to find out what errors are induced by her rigging and so on. But first you should warm up by taking a few practice bearings.

Most direction finding stations are on the air for only a few minutes at a time, so the navigator must tune the set to the station he has chosen—during a period when it is on—then tell the helmsman when it will come on again and give him the heading the boat should be on at that time. Then he sets the bearing ring on top of the set, while the helmsman gets the boat into position and takes up the heading.

When the station comes on, the navigator identifies it by its call sign, calls "Stand by" up the hatchway, and turns the antenna to find a null. And here comes the tricky part.

Let us assume there is virtually no signal for a width of fifteen degrees. The navigator turns the antenna to the left until he just hears it, then to the right until it sounds the same, and quickly jots down the two readings—say, 268° and 282°—then he calls out "Now." When the helmsman hears that, he reads the compass and calls back how far he was off the proper heading at that moment. So if his heading was 014 degrees and he was on 017 degrees, he calls "Plus three" and the navigator writes that down.

Now the mean of the two readings—which should be the middle of the null—was 275°, so the bearing, corrected for the helmsman's error but nothing else, was 278°.

Usually we take four such "shots" in quick succession, then add them together and divide by four to get one bearing. If one of the shots is obviously wild, discard that and try again, to get four reasonably good ones. So the navigator's scratch pad looks like this:

268 / 282 = 275 + 3 = 278
265 / 177 = 271 + 5 = 276
273 / 287 = 280 − 3 = 277
274 / 290 = 282 − 1 = 281
1112 by 4 = 278 degrees

This looks like a lot of work to get one bearing, but in fact it scarcely takes two minutes. In most cases it will be accurate to one or two degrees, and if it is not, the spread of the figures will warn you to be cautious in using it.

Now the deviation of a compass varies with the direction of the earth's magnetic field, relative to the boat, so that its deviation card can be written in terms of *north, south, east,* and *west.* But the deviation of a radio direction finder varies with the direction of the incoming signal, so that its deviation card has to be written in terms of *ahead, port bow, port beam, port quarter,* and so on. Therefore we first take a set of bearings with the boat heading straight toward the transmitting station, then one with her heading 45 degrees to the right, and so on. Thus, if the station bore 014 degrees (magnetic) we would take bearings with the boat heading 014, 059, 104, 149 degrees, and so on.

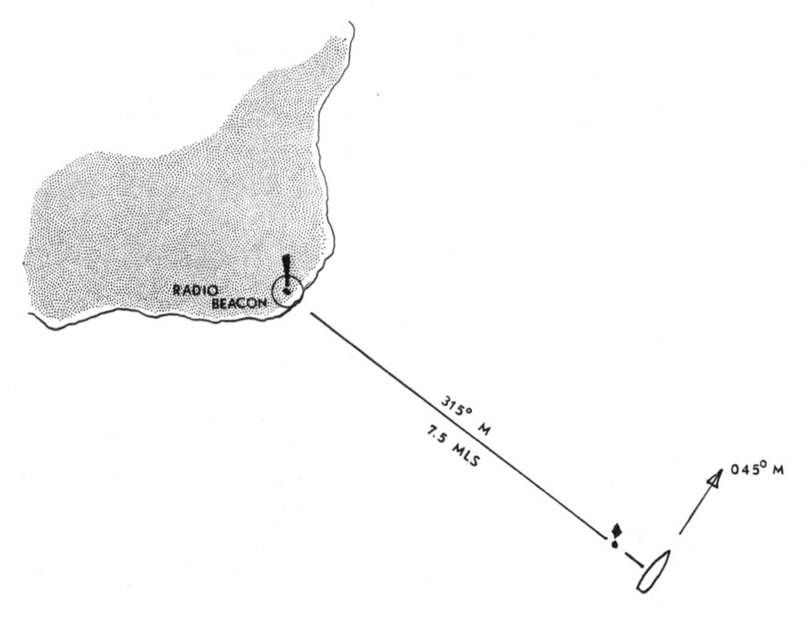

Checking the radio direction finder, on a heading that is 90° more than the bearing of the station. Such shots are taken at 45° intervals, all around the compass.

Once the bearings are taken, the helmsman can head for home, while the navigator makes out the radio direction finder's correction card, which is quite easy. He merely takes each bearing (which is the average of a set of four), corrects it for the deviation of the compass (for the heading he was on at the time), and compares it with the magnetic bearing from his position to the transmitting station. Any difference between the two is the direction finder's deviation for signals arriving from that particular direction, relative to the boat.

In a wood or plastic boat, that completes the job. For the deviation of the radio direction finder is seldom more than a few degrees on any heading and later, when you take a bearing for navigational purposes, you can interpolate between the 45-degree increments easily enough.

There is one other kind of error—caused by the heeling of the boat—but we find that can be ignored for practical purposes. So now you are all set. When you need a radio bearing, you take four shots, average them, apply one correction for the deviation of the boat's compass (depending on her heading) and another for the direction

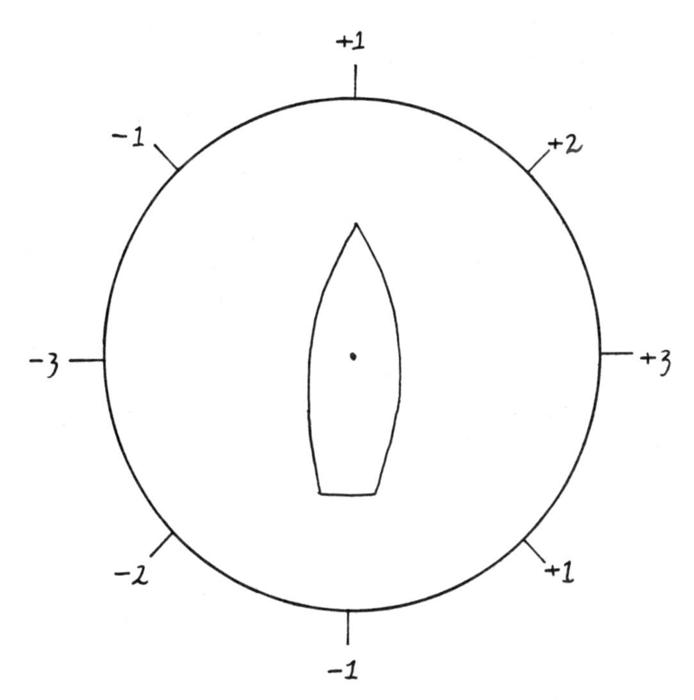

A correction card like this reminds the navigator that the deviation of the radio direction finder is a function of the direction from which the signal is arriving, relative to the boat, not relative to the bearing of the station.

finder (depending on the direction of the incoming signal, relative to the boat), and you have it.

Of course you must always watch out for signals that have crossed a coastline at an angle, but this problem seldom comes up. You must be quite firm about not taking bearings on any transmitter less than six miles away. So if you want to go right up to it, take a series of fixes before you get too close, turn off the set, and continue by dead reckoning.

In an aluminum boat a direction finder with its antenna below decks will not work, since the incoming signal is cut off by the hull. A set having a remote, hand-held antenna can be used. But you have to experiment with it to find a place on deck where you get satisfactory results and, once having established that, you must stick to it.

The same type of direction finder works well below decks in a wood or plastic boat and since it has a built-in compass, there is no need for the helmsman to read the boat's main compass each time the navigator takes a bearing. But in that case, you must find a place where its little compass is not affected by any stray magnetic fields within the boat and always make sure it is in exactly that place, each time you use it.

In a steel boat, the only satisfactory kind of direction finder is a remote system, having a pair of loop antennae up on a mast and the rest of it in the navigator's office. This works fairly well but is elaborate and expensive.

Automatic direction finders are useful in special cases —such as on a fishing boat that "homes in" to her local harbor at the end of the day—but they are not nearly as versatile as the other kinds and only work with strong, clear signals.

And it is surprising what you can do with an ordinary

set, once you get used to it. It is not hard to tell one station from another by ear and there are so many stations that a coastwise passage in fog is often scarcely more difficult than one on a clear night with lighthouses to guide you all the way.

3

Using a Storm Trysail

Many people have storm trysails aboard their boats, but hardly ever use them. This is a great shame, for they are most useful sails, especially in passage making, when you are often short handed and want to get somewhere, not as fast as possible but with the least wear and tear on the boat and her gear. Perhaps the word 'storm' is the problem. It would be easy to assume they were sails made for use only in storms. And they do look odd, as though they were designed for a special purpose. But in fact they are quite versatile.

Since the strength of the wind varies with the square of its speed, a thirty-knot wind has four times the power of a fifteen-knot one. And on a long passage, you must take what comes. So quite often, you either have too little wind or too much. When the wind falls light, you can always dig out extra sails and set them wherever they will go, to catch the slightest breeze. Becalmed in the Sargasso Sea, we once had every scrap of canvas aboard a large yawl up at the same time, including the dinghy's sail. But in heavy weather, you need only a couple of small sails and the question is which ones to use.

In those conditions you will usually be running, reaching, or hove-to and in each case a trysail has a lot to recommend it, compared with a double-reefed mainsail. For a start, it needs no boom, so you can secure that. And no longer need you be concerned about gybing, for with a trysail that simply isn't a problem. As the wind comes across your stern, it just flops over to the other side.

Being heavily made and having no battens, it can be left up as long as you like, in almost any weather—so long as you don't let it chafe on a shroud or a running backstay. In a strong wind it is a good driving sail and used with a working jib or staysail (as the rig of the boat dictates), it will give you plenty of speed with unique control.

If a ship suddenly appears out of the driving rain, dead ahead, you can change course without hesitation, to avoid her. You can run or reach, wherever you want to go, and even claw out to windward, if you keep her footing. And when you want to stop, you just put the helm down. As the boat comes about, the staysail goes aback and you are hove-to. In a moment, the fuss dies down and everything is peaceful. You can stratch your legs, lash the tiller, adjust the sheets, and go off to do other things.

So what is the snag? Getting the trysail up.

More often than not, by the time you decide to hand the reefed mainsail, the sea condition is such that working around the mast is less than pleasant. But if you are prepared for it, the change can be made without too much trouble.

We hope that you hoisted the trysail long ago, in a calm port, and showed your crew how it should be set. Its downhaul is in the bag with it and they know what sheets to use. Most trysails we have used had two sheets permanently attached to them but often those were too short, so we took them off and used the heaviest jib sheets available.

Unless there is a gallows for the main boom, plan on lowering its end to the deck, resting it on an old cushion, and securing it there. That gets it out of the way of the trysail and eliminates any chance of it falling on you later.

If you are lucky, there is a length of track on the mast to put the slides on, with a switch so that the mainsail can be lowered and the trysail hoisted right away. Otherwise, you take off the stop at the bottom of the main track, to let the mainsail's slides run off as you lower it, then feed the trysail's slides on at the same point. But either way, the idea is to get the trysail up as soon as possible, after the main comes down. So first you have everything ready and make sure that each man knows exactly what to do.

The proceedure varies, of course, with the size of the boat and her rig, but let us assume she is a forty-foot sloop. In that case, you have little choice of sails to steady her while you change over from the mainsail to the trysail, for the spitfire jib is too small to do the job and most Genoa jibs are too lightly built to use in heavy weather—which leaves the working jib. So you set that and sheet it hard in.

Then you bear off the wind, easing the main sheet until the boom is about half way out and the luff of the mainsail is lifting in the back wind from the jib. At this point, the boat will settle down and go slowly along, with comfortable steerage way but very little fuss. So the helmsman reads the compass and holds that heading, while the others take their places.

In a forty-footer on a passage, there may be no more than three people aboard, so the helmsman will have to handle the sheets and look after the main boom, while the other two stand by the mast, one to take down the mainsail and the other to slack off the halliard. But it should never be allowed to run free, or the reel may overrun the wire and throw a turn around the body of the winch. It should be payed out steadily, under enough tension to prevent it from blowing away from the mast and getting hooked around the end of a spreader.

With the boat on that heading, the sail will come down quite easily and as it does, the boom should be lowered to the deck and secured, which will give you something firm to hold onto while you furl the sail and stow it temporarily.

At that point, with the working jib set but nothing else, you will probably have to use lee helm to maintain your heading, but that is all right, for the boat will merely slow down and make more leeway, rolling a little but not too badly. Then the main halliard is transferred to the trysail and as one man puts the slides on the track, the other hauls it up.

There is no need to head into the wind, since the back wind from the jib will keep the slides running free on the track and the trysail will go up quite easily. But it is best to take it a foot or so too high, then make the halliard fast and use the downhaul to even the tension on the luff. Meanwhile, the helmsman takes in the sheet just enough to prevent the trysail from flogging, until he gets the signal to harden it in.

Then we like to put the boat on the other tack (to bring the boom onto the windward side) and heave to while we tidy up the lines and stow the mainsail properly. And that's it. We're all set now to go sailing under trysail and working jib.

In a smaller boat, there may be fewer people but her gear is so much lighter that it presents no problems; a larger boat usually has more men aboard, so the work can be spread out, with a man aft to handle the sheets and more to help with the sails. And the larger boats usually have other rigs, which give more flexibility.

In a cutter, you may have a Genoa staysail that goes very well with the trysail, while the working staysail is available as an alternate. But beware of the kind of staysail that has a boom, for they have a habit of sweeping everything off the foredeck in heavy weather, including people. In a yawl or a ketch, you can use the staysail and mizzen to steady the boat while you change the mainsail for the trysail. And in a schooner, you can use the staysail and foresail. So in each case, the problem of rolling is eliminated.

But there is one time when setting the trysail is easy: when you are starting out in heavy weather. Then you can hoist the sails in the calm water of the harbor, as you go out to sea. This works very well, sometimes enabling you to make a passage, which would otherwise be miserable, in comparative comfort.

And finally there is a compelling argument for using the trysail on a passage, whenever there is enough wind to make you wonder if you should: the cost of new sails. Most trysails die of old age, not overuse. There isn't a cat's chance in hell that you'll ever wear the thing out. And each time you use it, you are saving your mainsail, keeping it in good shape for the days when the wind will be blowing at the speeds it was designed for.

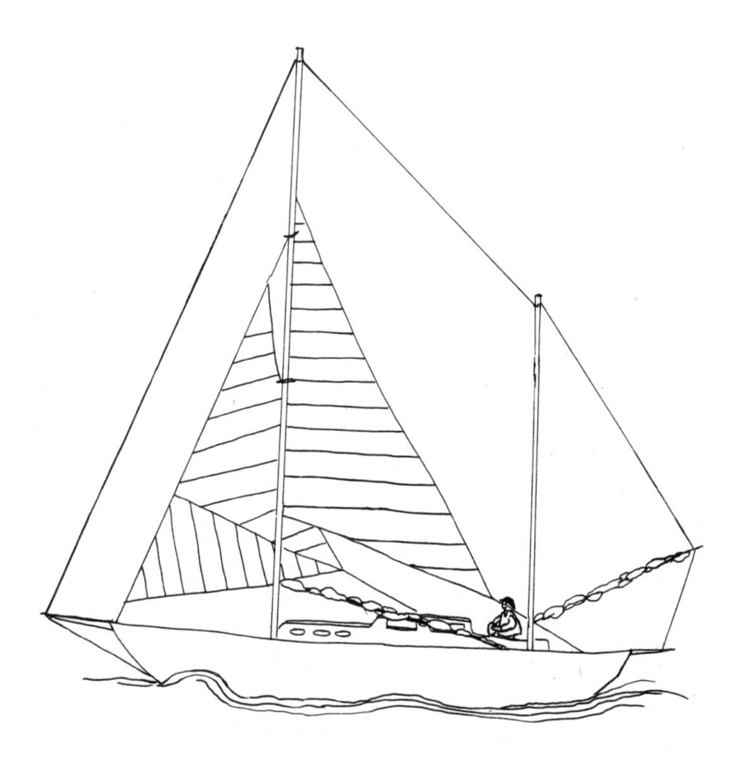

The yawl is reaching fast, in heavy weather, under Genoa staysail and storm trysail. The end of her main boom has been lowered and secured on deck. Her mainsail is still bent on but is tightly furled. And her mizzen boom is cocked up to clear the seas. The man in her cockpit is steering and keeping watch for other vessels —and for any change in the storm.

4

Navigation by Depth Finder

One evening we had to take a motorboat down Long Island Sound, to be at her destination in the morning. There was thick fog all the way and she had no navigational equipment aboard her, except a compass and a depth finder. Yet we made the overnight trip at a steady ten knots, without any trouble.

How? By using the depth finder to tell us where we were, how fast we were going, how much to correct our course for variations in the tidal currents, and exactly when to look out for the sea buoy at the other end.

The point is that a depth finder is far more than just a gadget to warn you against going aground. Properly used, it can often be a primary navigational instrument. But like all such instruments (including a sextant) it has to be studied and understood, its limitations appreciated, and the techniques for using it learned, before you can expect to get good results from it.

The most common type of depth finder is a smallish box with a large round dial on its face, usually marked in feet. Inside, you can see an arm that goes around when you turn it on. And at the end of the arm is a red light, which flashes in the same places each time it goes around.

The box is connected by a coaxial cable to a transducer in the hull of the boat below the water line. As the arm passes the zero mark, the light flashes and the transducer sends a pulse—a noise like 'tock'—down through the water. Then the instrument switches to "receive" and waits for the pulse to hit the bottom of the sea and bounce back up again.

As the pulse travels through the water, the arm continues to move around the dial. And when the echo of the pulse comes back to the transducer, the light flashes again. So the deeper the water is, the farther around the dial the arm will be when the light flashes for the second time. Since the cycle is repeated each time the arm goes around, you should see what appear to be two flashing lights on the dial, one at the zero mark and one at the depth of the sea.

But in real life, it is not quite that simple. The pulse may bounce back from anything it hits, such as a school of fish or a patch of bubbles in the water. It will come back more strongly from a hard surface like rock than from a soft one like mud. And you can often see several echoes on the dial at one time, at different depths. Some of those may be from fish or bubbles. Some may be transient signals, which flash once or twice and disappear. And quite a few may be from the bottom. For each pulse will hit the bottom in a number of different places and those echoes coming from points directly underneath the boat will obviously get back sooner than those that had farther to go.

So you have to interpret what you are looking at. And to do that, you first adjust the gain control (usually the same knob as the on-off switch). If you turn it up too high, you see lights flashing all around the dial. And if you turn it down too low, you lose all the echoes. But just above that level, there should be a point where you can make out what is going on.

If the bottom is hard and not very far down, it will show up clearly as a group of echoes, close together on the dial and stronger and steadier than any others. And in that case, the depth of the water is the least of those.

But if the bottom is soft or the water is deep, you may have to turn the gain up so far to see it that the dial becomes cluttered with other echoes. There may even be some doubt as to which group of them really is the bottom. In that case you look at the chart, find your position by dead reckoning, and see what it says there. Not only will it give you the approximate depth of the water but also the type of bottom. With that information you can go back to the depth finder and identify the group of echoes you want. If there is another group of echoes, like those from the bottom but weaker and less deep, it may be a school of fish or the wake of a ship. But keep an eye on it, anyway.

After using a depth finder of that kind for a while, it is possible to get quite good at reading the whole picture on the dial—for example: The bottom is at eighty-seven feet and is hard. There is a school of fish at forty feet and the wake of a ship at fifteen feet but that seems to be dissipating. If you read it at regular intervals on a passage, you will notice changes in the depth and the type of bottom, which you can compare with a chart. And quite often you find features that fix your position.

But first we should check the depth finder out. Hopefully, it was installed by someone who understood its requirements. It is a good idea to read the instructions that came with it and make sure that the transducer is in a suitable place (pointing straight down if there is only one or angled outward if there are two), and away from any turbulence in the water. When the boat is hauled, you should also make sure that there are no barnacles on the face of the transducer, or just forward of it, where their wake could flow across it.

The instrument itself should be mounted over the chart table, where it is out of the weather and the navigator can use it. And we prefer to adjust the zero setting to allow for the depth of the transducer below the water line, so that in effect it measures downward from the surface of the water.

Then we take the boat out and check the accuracy of the depth finder against a lead line. If it is right at two or three depths, we are in business. Otherwise, it has to go to a service man who has the equipment to correct it.

In use, a depth finder of that kind has some limitations. It needs electricity to run on. It will not work at high speeds, because of turbulence around the transducer. It may not "see" the bottom at large, sustained angles of heel, because the beam of the downward pulse is relatively narrow. It will not stop you from going aground in shallow water, because that happens too quickly. And in very deep water, the bottom is out of its reach.

But on a coastwise passage in fog, when you are going at moderate speeds and not heeling a great deal, it can be a very useful navigational instrument. Any contour line on a chart is a ready-made position line that you don't even

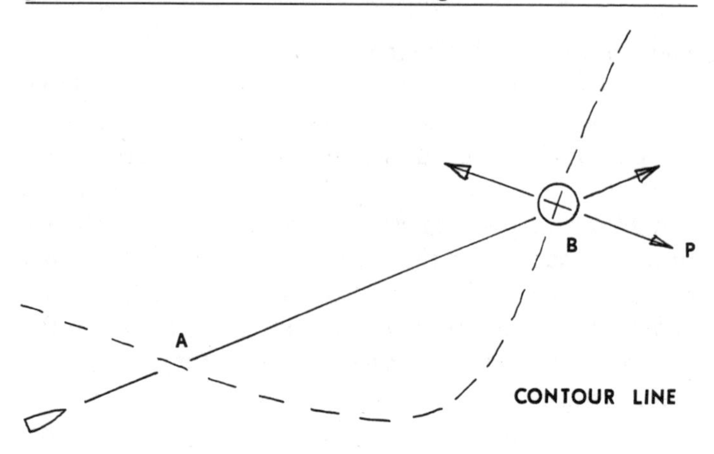

A fix by one contour line. The contour line at point A can be carried forward, to become a position line (P) that intersects with the contour at point B to fix the position of the boat.

have to draw in. All you need do is look up the state of the tide, allow for that, and note the time when the depth finder says you are there. Like any position line, it can be advanced with the boat. So if you can find a second one ahead that will cross it at a reasonable angle, you have the makings of a fix.

Since contour lines often bend around quite sharply, you may be able to use the same one twice. That happened that night on Long Island Sound. Halfway down, we crossed a V-shaped trench and used the same contour line on each side of it to get a fix, from which we figured the corrections to make in our heading and time of arrival.

In other places, where the contour lines are long and straight, following one will often take you safely down the coast, until you come to a feature that gives you a fix. The feature may be a gully cut by a small river flowing out to sea, or a pile of rocks (safely below the surface), or anything else that you can positively identify. And since the contour lines and features are all on the charts, you can decide before you sail whether there will be enough information available for you to make the passage safely in fog, using the depth finder as your primary navigational instrument.

Most kinds of depth finder use the same principle (known as echo sounding) but some present the information in different ways. The digital type is easy to read, while the meter type uses very little current and can run off dry batteries, so that it won't fail if the boat's electrical system does. But in both cases, you only get one reading —one you have to hope represents the bottom and not something else.

Some boats have twin transducers, angled outward on each side, to take care of large angles of heel. Others have special fairings in way of their transducers, so that soundings can be taken at higher speeds. And in a large yacht, the transducer is sometimes in a water-filled cofferdam inside the hull.

Commercial vessels often have recording depth finders, that are similar to the kind we described, except that instead of a flashing light, they have a pen that goes around, tracing the information on a roll of paper. Those are very good, since they give you all the facts and don't let you forget them. But they are larger and more expensive and they sometimes run out of paper at awkward moments.

We have found the ordinary kind of depth finder to be so reliable that we could hardly ask for more. Except of course a radio direction finder to use with it.

When both of those instruments are properly installed and calibrated, navigation in fog can be a pleasure. On

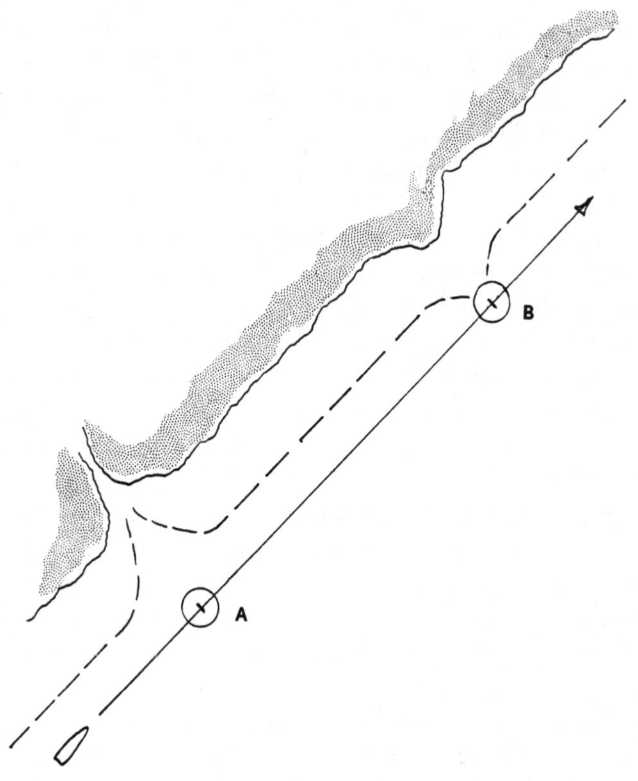

Boat is running along a shore in fog, in about 15 fathoms of water, and anomalies—deepening at point A by a river and shoaling at point B off a headland—fix her position.

deck, the crew may not be able to see ten yards, but below the navigator often has more information than he needs.

Since the depth finder is available all the time—once you are on soundings—while the direction-finding stations come on at different times, we schedule our work to suit those. As each one comes on, we take its bearing, then record the depth of water (corrected for the state of the

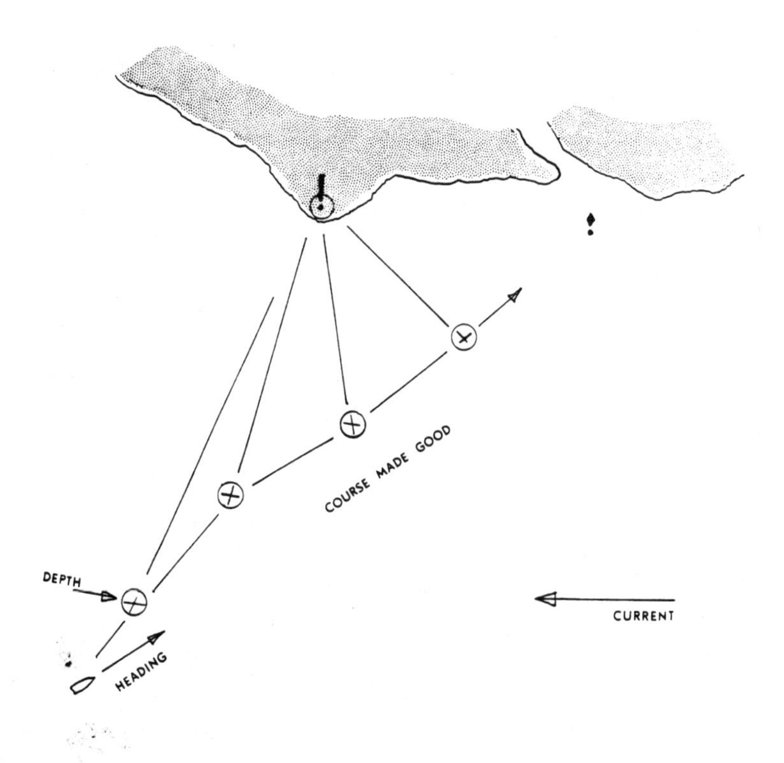

A series of fixes—taken with the radio direction finder and the depth finder—lets the navigator correct his course in time to pick up a buoy. The boat's track is over the ground and her heading is adjusted to allow for the current.

tide) and the time.

Quite often, one bearing and the depth that goes with it will give you a fix, so you plot that. Otherwise, the bearing is a position line, to be advanced and used with one from another station, or perhaps with a contour line. So you fetch up with a whole string of plots, marching steadily across the chart.

That kind of navigation is over the ground, so we use the time as our reference and the patent log—if we have one—to check the current. And by drawing a line down the middle of the plots, then measuring the distances between them, we find that we can determine our position, course, and speed with sufficient accuracy to pick up a buoy in nearly zero visibility.

How can you learn to do the same? There is only one way, really, and that is by practicing in your boat, at sea. You don't have to be in fog the first time. Try it on a fine day, with the crew on deck looking out but not telling you what they see until it is within fifty feet or so.

And don't cheat. Cover the cabin windows and stay below. For in fog, there will be nothing to look at outside and below, there is plenty to do: Using the instruments, interpreting what they tell you, making corrections and plotting the results. But after a while, you will find you can see the whole situation on the chart. And then you can direct the boat from where you are, without feeling any need to go on deck.

5

Witching and Dredging

In the old days, when a sailing boat often had no engine, there were tricks one learned for handling her, like witching along a bank or dredging an anchor into a berth. And though you seldom need these tricks today, they are good to know when you do.

Witching is a procedure for sailing to windward through a land cut (provided it has deep water up to its banks), which is not wide enough to beat through in the normal way.

As you approach the land cut, check the direction of the wind, for though it often blows almost straight down a cut, it seldom does so exactly and there is usually one tack which is more favorable than the other. So you enter the cut on that tack, with plenty of way on the boat, and as you close the far bank, you head your boat up (trimming the sheets as you go) until she is parallel with the shore, a few feet away. And she will slide along—almost dead into the wind—for a hundred yards or so, before she loses headway and you have to go about.

The explanation is that the water, trapped between her bow and the bank, exerts a force which prevents her from making any leeway. So she goes straight along, like a very slow ice boat. But when she loses headway, the force decreases rapidly, until suddenly you realize you have had it.

The other tack is the bad one, so you make no attempt to use the opposite bank but concentrate on gathering all the speed you can, as you go across the cut and back. Then you round up into the wind and go witching along the shore again.

Of course, if the cut has a bend in it, you may have to switch sides, as the other one becomes more favorable. And you can use the same procedure to work along the bank of a wide river, when there is a current against you (going out just far enough to gather the headway you need, then dodging back into the slack water, along the shore). But once you understand the principle, you can use your own variations.

Dredging an anchor is a means of taking a boat sideways into a tight berth (as alongside a quay), using the current as her motive power, with positive control all the way.

In the drawing, there is a gap between boats A and B that you wish to occupy. The current is running down the river, as indicated by the arrow and the wind is coming from the same direction, more or less. So you beat up river to position C, where you round up into the wind, drop the anchor, and furl the sails, which leaves you lying to the current in position D.

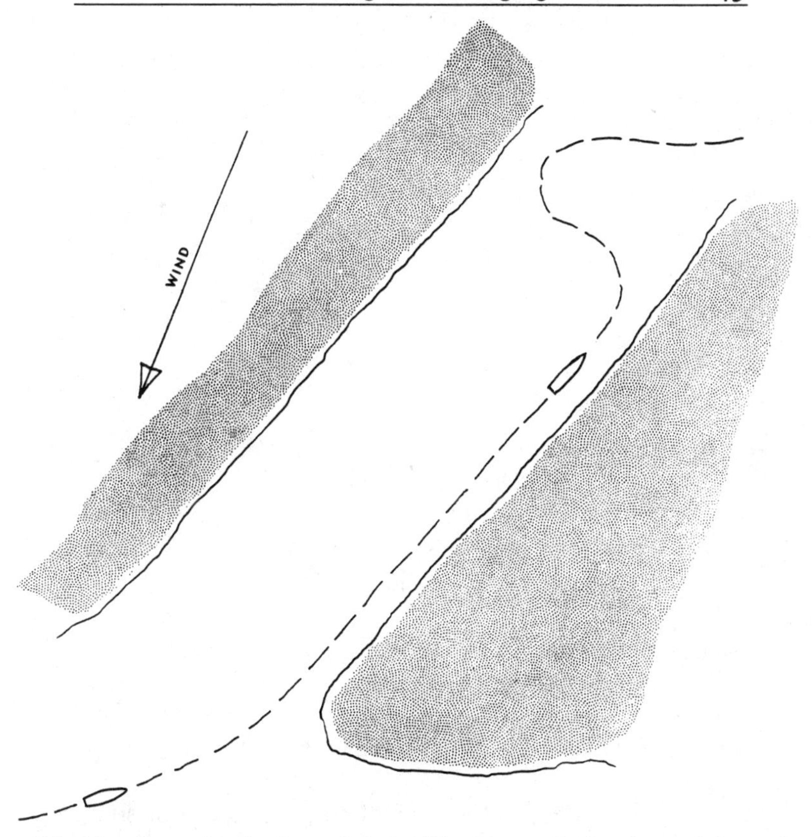

Witching through a land cut that would otherwise be hard to sail through, in a wind from the direction shown.

Now you put the helm over to starboard (or the wheel to port) and the boat will take a sheer toward the quay, until she is lying at an angle to the current, with the anchor rode leading off her starboard bow, in position E.

Then you take in the rode, a little at a time, until the anchor begins to drag (or dredge) slowly along the bottom.

As it does, her bow will fall off a little, so the angle

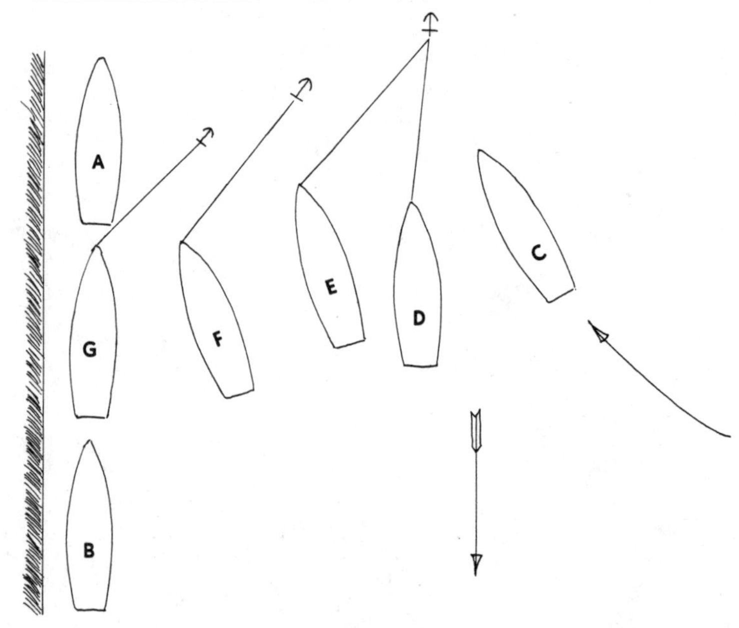

Dredging into a berth. Boats A and B are lying alongside a quay. The current is running in the direction shown by the arrow and the wind is blowing approximately the same way. Our boat goes from position C to position G—as described in the text.

between the rode and the current will increase. But since the anchor is being dragged by the rode, the boat will proceed in the direction that the rode is pointing, which in this case is toward position F.

The advantage of dredging is that it gives you such good control. If the boat stops moving, you take in some rode and she starts again. If she goes too fast, you pay out more rode, until she slows down—or stops altogether, if you wish.

If the boat is not going in the right direction, you can correct it by changing the rudder angle. And if you want to discontinue the maneuver, you simply put the rudder over the other way and she will sheer back out into the river. So it is quite easy to bring her slowly and safely into position G alongside the quay, even when there is little room to spare.

And when the boat is made fast, it is not hard to get the anchor up, for the scope of the rode decreases as you go.

Dredging can also be used to move a boat out of danger, when no other means of propulsion is available. For example, one night we were aboard a small motorboat with a dead engine in the Seine River, just below Rouen. It was very dark. When the motor quit, her lights went out, so the ships coming up the channel would never see her. And as she drifted to a stop, her rudder went dead. The night was calm, with not a breath of wind, and soon she lay quite still on the water. The engine refused to start, of course, and searching below, we found there was not even a paddle to move her with. But the ebb tide was running gently down the river. So we lashed the wheel hard over, then went forward and dropped the anchor. The moment it caught, the boat came alive. Ripples of water chuckled past her hull and she took a sheer toward a bank. As we hauled in the rode, the anchor started dredging and soon we were safely out of the channel.

So it is a good idea to practice witching and dredging, when you get the chance. Then if you need to use one of them, it will not be the first time you have done it.

6

Understanding the Weather

Nothing adds more to the safety and comfort of a passage than a thorough understanding of the weather, as it affects the sea condition and the performance of your boat.

For a day's run down the coast, it enables you to say in advance, with a fair degree of accuracy, what time you will leave, what courses you will sail, what will happen along the way, and when you will arrive at your destination. In a powerboat, that might mean arriving within a few minutes of your estimated time. Even under sail, you should hardly be off by more than ten minutes.

For an offshore passage, it helps you select the best route and the best time to start out, then to decide as you go along when to cook a large meal, when to prepare for heavy weather, when to work on the engine, and when to let your crew rest, so as to achieve the maximum efficiency. And even on an intracoastal run, it lets you adjust your timing so as to avoid such foolishness as fueling in the rain or docking in the middle of a thunderstorm.

But like all good things, it has its price.

First you have to study basic meterology, so as to understand the different types of air masses and their behaviour, the various kinds of weather system and how they develop. And though you do not have to memorize the intricate details of the craft, you need a solid grasp of the fundamentals. Then you can begin to relate the movements of the large systems to your own local observations, to see for yourself what barometric changes, cloud formations, temperature changes, wind directions and forces are associated with the passage of each kind as it approaches, passes overhead, and goes away.

A convenient way to do that is to take a newspaper like the *New York Times,* which has a good weather map, and compare it with a log that you keep. All you need is a barometer, a note book, and a place where you can see the sky—the roof of a building will do nicely, or a big parking lot.

On a clear day, you can get your bearings by the sun and make a few discrete marks from which to tell the wind direction, while the wind forces are admirably described in the Beaufort Scale. And we have always found the forces far easier to relate to sea conditions than speeds in knots.

The clouds we record by type, in tenths of the sky, as: "4/10 cumulus" or "2/10 cirrus." The visibility is not hard to estimate, in an area you know. And the temperature or humidity are not critical, so that "cooler" or "dry" will do for those. But the barometric pressure is important.

Most barometers have a needle you can set, to tell you

where it was the last time you read it. But we find this needle only gets in the way, so we turn it all the way down. Then we tap the instrument lightly to help it make its mind up and read what it says, to the nearest hundredth of an inch. After which, we look in our log to see if it is rising or falling.

A set of observations like that only takes a few minutes and if you do it two or three times a day, you will soon get the feel of the different weather systems, so that a cold front on the map becomes a thing you can see and feel and hear, on the roof, with a character of its own that is quite unique.

Then you can start making your own predictions from the map and see how they turn out in real life. Take a front that is several hundred miles away and try to figure when it will get to you, how it will change before it does, what part of it will pass over you, and what it will be like. How will the clouds look, as it approaches? What will the barometer say? What will be the direction and force of the wind? The visibility? And so on, at each stage of its progress.

If you are right, first time, it was a fluke. It is not that easy. But it comes with practice and the results get better all the time, as you learn to refine your estimates until you can often do better than a forecaster. This is not as outrageous as it sounds, for he has to cover a large area and satisfy the demands of farmers, airplane pilots, hamburger stand operators, and God knows who else, while you only care about what happens in one place, from one very narrow point of view: How will it affect your boat?

Precipitation, which seems to bother people most, is irrelevant, except as it affects the visibility. Rain is just a nuisance. Temperature and humidity are only of interest to help you tell what is going on. The thing that really matters is the surface wind, for that determines the sea condition.

As the air flows over the surface of the water, it makes waves, which slowly build up into seas. In shoal water, that may take two or three hours, while off soundings we allow five hours for them to reach their full size, which depends primarily on the depth of the water, the force of the wind, and how far it has come from the shore (called its *fetch*).

In shallow water, friction between the sea and the bottom absorbs the energy of the waves and limits their size. Of course, where a heavy sea has already developed, it will pile up and break on a steep-to shore, before the friction has time to absorb all that energy. But a gently sloping shore will gradually reduce the size of the waves. With a steady wind and a level bottom, the waves will stop getting bigger when the energy they are losing through friction equals the energy they are getting from the wind.

Since the waves are moving along as they get bigger, it takes a good fetch of deep water to build a heavy sea, though once it gets started, it will keep on going, long after the wind that created it has changed direction or moved away. So as a weather system moves across the ocean, you will often find a new sea building, at right angles to an old swell or even in the opposite direction, with the new waves running over the old ones. And in a calm, you are usually rolling on the swell left over from the last wind. But if a new swell comes in, without any wind to account for it, watch out. For whatever made it is probably large and powerful and coming your way.

Far out in the ocean, the length of a sea is about the square of its height in feet. So a fifteen-foot sea should be over two hundred feet long and a twenty-footer would be more like four hundred feet, from crest to crest. And those hardly bother you, unless they are breaking, since you rise so gently over them. But inshore they are often steeper and there you can run into a head sea, eight feet high and the length of your boat, that will really stop you. So when you are estimating the effect they will have on your speed, you have to consider the shape of the seas, as well as their height.

Once you understand what is going on, you can study the effect of the wind on the sea condition and of that on the speed of your boat, each time you go out in her. But write down what you find out, before you forget it. And soon you can start predicting exactly how a short passage will go.

To do that, you first gather all the information about the weather that you can, from the map in the newspaper, the reports on TV and your own observations. Where they are available, the radio reports on VHF–FM are also helpful but those on the commercial broadcast stations are seldom worth waiting for, so it is better to phone the nearest weather bureau and listen to their recorded announcement for the latest details.

Next you make your own forecast of the direction and force of the wind, at various points along your proposed route, at the times you expect to be there. And relating those to the depths of water and bottom configurations, shown on the charts, you can estimate the sea condition at each point.

That enables you to figure what courses and speeds will be available to you and to select those which seem best. Then you can plot the whole passage on your charts—

with the courses, speeds and times for each leg—and decide on the best times for your departure and arrival.

Finally, you check the whole thing to make sure that no errors have crept in. Will the wind and sea be the same, at each point on your final schedule, as in your first estimates? Have you shown a leg that can not really be sailed in the time you allowed for it? Are all the currents figured in? What will the visibility be like, when you need it? And so on.

After this comes the proof. You make the passage, keep a detailed log, and see how it compares with your predictions.

Once again, you should not be disappointed if you don't do too well, the first time. It is not that easy. But each time you make an error and see what it is, you learn something. And in due course, you should get pretty good at it.

Meanwhile you can fool your crew by adding ten minutes to your estimated time of arrival. That is purely psychological but it works. For if you are ten minutes late, they think you are a bum. They may even jeer. But being ten minutes early is all right, and being on time is even better. So you can hardly lose.

On longer coastwise passages, the errors you make tend to cancel themselves out, which helps offset the greater difficulty of estimating how the weather situation will develop. And as you progress, you can take more things into account.

Where a current runs against the wind, the seas will be steeper than usual, and vice versa. Inshore, a strong wind can upset the normal ebb and flow of the tide. Offshore, a steady wind in one direction may set up a surface drift in the water. The land and sea breezes add to or subtract from the wind, near the coast. And so on. But the more you see of such phenomena, the easier it becomes to predict them.

On an offshore passage you can predict the weather for the first few days only, so you have to play the averages, which are shown in the Pilot Charts. The first step is to plan the trip on those, selecting the best route (not always the obvious one) and the best month in which to sail.

When all is ready, you gather your weather information in the usual way and make your own forecast. If it looks bad, you stay in port, for it would be stupid to start out in bad weather and get everyone's stomach upset, before they had time to settle down. But if it looks good, you phone the nearest weather bureau and speak to the meterologist on duty.

We never bother those people except for a serious passage but we have always found them most helpful when we do. First of all, you want to know what is going on. Are the systems that are coming your way developing as you expected? And are they still moving at the same speed? Then you would like an extended forecast and if that says "Go," you go.

Some yachts have radios that can get weather information at great distances and as long as they work, we use them. But of all the offshore passages we have made, about half have finished with the boat's electrical system out. So you really need to be able to get along without radio. And our favorite substitute is a piece of paper with quarter inch squares on it.

On that we keep a graph of the barometric pressure, with a square (up and down) for a tenth of an inch of mercury and a square (across) for eight hours, which gives us a clear picture of each weather system as it passes over

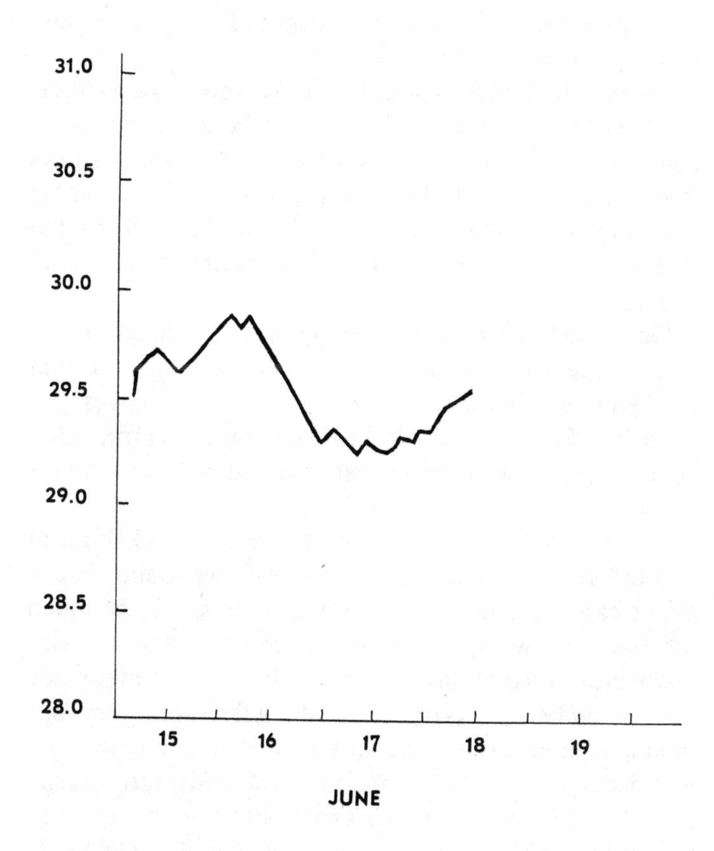

A graph of the barometric pressure, made like this, gives a much clearer picture than you get from a barograph.

us. That is better than a barograph and will show the diurnal variation in the tropics, but with such a high scale as that, you should consult the Pilot Chart to see how the mean pressure varies over the length of the passage and then take that into account.

Every four hours, when you make your observations, consider the big picture. What must be going on to account for what you have recorded? For the first few days it is easy to figure out, because you know what is coming your way. And later you get into the rhythm of it, so that you can still arrive at a reasonable estimate of the overall situation.

Once you know that, you can say what the direction and force of the wind will be for the next four hours and from that you can figure the sea condition, which gives you enough information to decide what sail to carry, what course to steer, what to do that day, and what to have for dinner.

There have been times when the data we had could fit two different weather pictures but then we found that in either case the wind direction and force would be much the same, so we split the difference and came up with satisfactory predictions. In fact, we have sometimes made passages of two weeks or more, calling the shots every four hours, without ever being so far off that any correction was necessary. At the other end of the scale, intracoastal passages are almost weather proof but once you are used to keeping an eye on the sky, you instinctively avoid silly situations. And being aware of what is going on above and around you makes even the most ordinary trip a great deal more interesting.

7

Choosing a Boat

There is no ideal boat. Each one is a compromise, in her design, construction, and equipment, intended to meet—as nearly as possible—the needs of a person or group of people.

In the first category are the custom boats, designed and built for their owners. If you know exactly what you want, such a boat can be excellent. In fact, it may be better to have a small boat like that, than a large one built for someone else.

To get such a boat, you start by selecting a designer. A good way of doing that is to look at many boats, by reputable naval architects, until you find out who designs ones that have most of the features you consider to be important. For a racing boat, you need a designer of winners, but for a cruising yacht—sail or power—we prefer more conservative designers, whose boats are sea kindly and carefully thought out, giving long service in all weathers with minimal trouble.

Explain to the designer exactly how and where the boat will be used—what waters you intend to sail in her, with how many people aboard, at what times of the year, and so forth—so that he can make the compromises that suit you best. When the preliminary drawings are done, you have a chance to make changes, if he did not understand your needs or perhaps has different ideas on certain aspects of the design. Then come the final drawings and the selection of a builder.

Ideally, you want a builder who has built other boats like yours and gets along well with your designer, whose prices are in line with your budget, and whose yard is near enough for you to visit, a couple of times a week, as the work progresses. Usually you have to compromise, settling for those things that are most important to you. Frequent visits to the yard are high on our list, because they enable you to catch any small errors while they can still be corrected easily.

The best engines for cruising boats are the ones that have been around forever, for the bugs are out of them, you can get parts, and people know how to fix them. Except in cases where weight is critical, we would always choose diesels over gasoline engines, because of the greatly reduced fire hazard, the greater reliability, and the much lower fuel consumption.

In a sailboat, the best size of engine depends on how she will be used. For weekend sailing—where getting back on Monday morning is important—an engine big enough to punch her into a head sea (with a reduction gear and a big propeller) makes sense. For long-distance cruising, we prefer a much smaller engine—big enough to move her sedately along in calm water—because it gives us the greatest range, per pound of fuel.

In a power boat the choice of engines is an integral part of her design, but it is best not to ask for more speed than you really need, for a slower boat is generally more comfortable and seaworthy, as well as more economical to run.

Before the boat is launched or rigged, you may want to install the sensors for some instruments: a through-hull fitting for the depth finder (or two, angled outward, in a sailboat) and perhaps one for a speedometer, a wind speed sensor on a mast, and maybe a wind direction sensor up there, as well.

For most boats, a depth finder is too useful to forgo. But for cruising purposes, a speedometer is of doubtful value —even with a patent log feature—because its underwater unit is too easily fouled, either by seaweed or barnacles.

Those units come in three kinds: a tiny propeller with its axis fore and aft, a paddle wheel with its axis athwartships, or a "finger" hinged to swing aft and inboard, against a spring, as the pressure of the water passing it increases. The first two measure speed by the frequency of electrical pulses from the sensor and distance by counting them. But in the third kind, speed is derived from the angle of the finger, while the distance is computed by integration against time.

The first two kinds are susceptible to fouling by the fine weed—almost as thin as a hair—that you find now and then in the sea. And the third kind often gets fouled, after a while, by barnacles on the hull, lodged against the finger (which make it stick at one speed) or by barnacles on the finger itself (which make it read high, in speed and therefore distance).

So we prefer to use the old-fashioned kind of patent log —a simple, mechanical instrument on the taffrail, with a line to a spinner in the water, behind the boat—to find our distance run and then we figure our speed from how long it took. A shark takes a spinner once in a while (though we paint them black, to prevent them from flashing in the sunlight), but we carry spares. And when one is fouled by weed, it is easy enough to clear.

For a sailboat, it is nice to have indicators for both the speed and direction of the wind, while in a powerboat the speed alone is useful, for keeping an eye on the weather.

When you buy a stock boat, such things are options and you can see from the start what you are getting. But stock boats are designed to meet the needs of many people—not just yours—and you have to consider the different features of the various kinds available, to arrive at the one that (notwithstanding the claims of the salesmen) comes closest to your requirements.

Boats depreciate rapidly at first, so a used one—two or three years old—can be a good buy, if she has been well cared for. An economical way to get one is to decide exactly what you want, then go around to several yards and brokers at the end of the season, leaving your name and asking them to call you when they are offered such a boat in trade for another one. That way the dealer makes his profit on the new boat, without being left with the used one, so you should get her for a little more than he allowed the previous owner, who traded her in.

In fact we often advise a person to buy a used boat if he is getting into a new situation—a new area, a new kind of boat, or one of quite a different size—to use for a year or so, then sell when he finds out what he really wants. But in that case he needs one he can sell easily, which means one of a kind that are popular among the people in the surrounding area.

Of course you must be careful when buying a used boat.

Any offer you make should be subject to survey and before you close a deal, you should get all the information about her that you can, as well as hiring the best surveyor around.

And we would never buy a winning racer, because her price would probably reflect her success—which is more likely to be due to her skipper and crew than anything else—while her hull and gear may well be much worn, from the hard use.

Nor would we buy a boat more than three years old, unless she were very simple, because you can find a rat's nest of dead wires and a warren of pipes that lead nowhere in a very old one. You may never untangle the mess during your ownership.

Apart from depreciation, much of the cost of owning a boat is for maintenance and that is a function of three main factors: size (the cube of her length), complexity, and age. So a big, old boat with all kinds of equipment is about the worst possible buy and a smaller, newer, simpler one is likely to cost you less, in the long run, even if you have to pay more for her.

Of stock boats, those that are better built—and usually cost more, in the beginning—are likely to hold up better and need less maintenance than their cheaper counterparts, provided of course that their equipment is no more complex.

And finally there are special situations that may give you an opportunity to buy a good boat for a low price but those tend to occur in far places, such as Portugal and Tahiti.

Each year, a number of people set out to make long voyages in boats, usually departing from Europe or the United States and heading westward around the world.

But some of them get together and share the cost of a boat that is jointly owned. And then it is common for them to split up, when they get to the first port, since friction has developed among the personalities. As soon as one member of the group demands his investment back, the others may be forced to put the boat up for sale, as they do not have enough money (after the inevitable overruns at the start) to buy him out. And there she stays.

So if you stroll along the waterfront in Lisbon or Papeete and ask a few questions, you may find a good boat, well equipped for ocean voyaging, available at a very fair price. But we would not suggest you pay the fare to go and look there.

8

Sailing Off an Anchor

Let us assume that you are sailing alone, along a rocky coast, in a thirty-foot sloop. It is summer and the barometer is high, so you anchor for the night in an open bay, rather than go far out of your way to a more secure place. As you put up the anchor light, you look around the bay. From north to south through the west it is sheltered by cliffs, rising up to trees whose tops move gently in the southwesterly wind. But from the northeast to the southeast it is wide open, clear to the far horizon. When you turn in, it is flat calm and everything looks good for the night, so you sleep soundly. Until you wake up to find the boat bouncing and tugging at her anchor.

On deck, you discover the wind has gone into the northeast and is piping up. Already the seas are building from the new direction, rolling into the bay, lifting the boat, and breaking on the rocks that are now behind her. Clearly you have got to get out of there, while you still can.

You put on your foul weather gear and try to start the

engine. But it won't even turn over. The battery is flat, which means you have no radio transmitter, either. As your eyes get used to the dark, you can see the rocks are so close that even if you could get the anchor up without the engine to move her ahead, the boat will be blown back onto them before you can set the sails.

On the face of it, you are in a tricky situation. But in fact, there is a procedure that will get you out of it, called *sailing off the anchor.* Here's how it goes.

First you check the wind direction. The best way to do that in the dark is to sight over the main compass, facing the wind and turning your head from side to side until you feel the same amount on each ear. Then you look at the chart and decide which tack to start off on.

Next you take the anchor light off the forestay, hank on the working jib, and get it ready to go up on the proper side for the tack you have chosen. Secure it with a single stop, using a slip knot that you can release in a moment.

If the helm was lashed, clear it. Then hoist the mainsail (reefed, if necessary) and sheet it hard in.

With the tiller swinging free, the boat will take a sheer to one side and as the sail fills, she will go ahead for a few yards. Then the rode will check her, she will round up into the wind, and go off on the other tack. All by herself.

Meanwhile you go forward and start taking in the anchor rode. It is really quite easy, if you time it right. Each time she rounds up, the rode goes slack for a moment, so you gather some in, make it fast, and wait for the next time.

That works you straight out to windward, as you shorten the rode, with little effort. Until it is up and down.

Now comes the only delicate part. The boat will break the anchor out for you, as she overruns it. But don't let

Sailing off an anchor. Taking in the anchor rode each time the boat heads up into the wind, with the mainsail set and the jib ready to go up as soon as the boat breaks the anchor out.

her do it on the wrong tack. Play it cool, wait until she rounds up to go onto the tack you want, then take the line in fast and snub it. As she breaks the anchor out, you haul it up quickly, secure it temporarily, and get the jib up.

With the anchor up, the boat will continue to fall off the wind as you hoist the sail, so you nip back aft, trim the jib sheet, ease the main a little to get her going, and beat out of the bay, into open water. As easily as that.

Who worked out that proceedure? We don't know. It must go way back, though, for it was common practice among the fishermen we sailed with forty years ago, who had no engines. And it's a nice thing to have up your sleeve when you need it.

9

Living Aboard

For a few days, as many people can live in a boat as there are bunks in her. But for a vacation of two or three weeks, people bring more stuff, and if they are crowded, they may get on each other's nerves. So it is better to have fewer of them. And on a long cruise, of several months' duration, it is best not to have any more people aboard than you absolutely need.

If the boat were under way all the time—as in a race, or on a delivery passage—crowding would be less of a problem, for some people would be on deck at any given time. But on a cruise, where you may be in port for days on end, you need more space to move around and to stow your "shore-going" gear. The best number of people for a long cruise is two and the next best is four (three is the worst number) and it is good to have as large a boat as you can safely handle in any weather, so that you have as much space per person as possible.

On a long cruise, the boat herself requires more space

for the stowage of tools, spare parts, and other things that you may not be able to get along the way. And you may need to take some special items, for unusually hot or cold weather.

In planning a long cruise, it is very important to stay in the best weather you can. For example, a boat will leave Europe in the fall, before the winter gales start, and head south, to the Canary Islands, then cross the Atlantic Ocean in mid-winter (when the chance of a hurricane is least), and go north, with the sun, through the Caribbean, and up to New York. That way, she can make good passages—with the least wear and tear—while her crew enjoy good conditions, all of the way. But it is not always possible to plan a cruise so well. And then you may have to wait out a period of bad weather.

One such situation occurs if you have to stay in the Caribbean through a hurricane season, before going on to the Pacific. Then the trick is to find a place where your boat will be safe—even if a hurricane hits you fair and square—and be there before the season begins, which is usually toward the end of July.

As a hurricane approaches a shore, the water level may go up as much as twelve feet and it is important to secure the boat so that she can rise that much, then come back down—as it passes—without being stranded on anything like a dock.

You also need shelter from wind and especially from waves, which means being in a small, enclosed area of water, preferably with a bottom of soft mud. A narrow, winding river with trees on each side of it, as far from the sea as you can get, is a good bet. There you run lines to the trees, so as to keep the boat clear of all dangers, somewhere in the middle. The anchor rode will do for two of the lines but you need a couple more. Two long warps, which can also be used as docking lines in other situations, are just right.

In the hurricane season, the temperatures are highest and there may be little breeze in a sheltered place, so awnings are essential, not only over the cockpit but also to shade the deck and thereby reduce the heat below. The best kind are those that have battens running athwartships every few feet and lines from the ends of the battens down to the rail, because they provide a good shade and leave the decks clear to walk on.

Another problem in that kind of weather is mildew. We once put a pair of polished leather shoes in a locker and by the next day, they were green. So you have to be sure that all of the enclosed spaces in the boat are well ventilated. It never pays to leave anything, like clothing, damp for long.

Similar problems arise when you have to stay in a northern, or far southern, port during a winter season. There you also have to find shelter from gales. When you live aboard, condensation underneath the deck creates damp in the lockers. But if you have electricity from a shore line, a couple of heaters with fans in them will usually dry things out and bring the cabin temperature up to a comfortable level, even with ice outside. But beware of heaters that burn kerosene or similar fuels, because they may create damp from the combustion of oxygen with hydrogen. And if you do not provide enough air for them, you can die from asphyxiation while asleep in your bunk.

Snow on the decks tends to make it warmer below, by giving you some insulation. But ice on a deck is extremely slippery and if you fall overboard, your survival time may be very short. The ice may damage the boat's hull, unless you keep it broken up for a foot or two, all around the boat, every day. A metal or plastic boat can proceed through thin ice but a wooden one is liable to be cut clear through—at the water line—in a short time.

Cold weather can have interesting effects. One dark night, we ran a water hose across a road to a tap on the other side and as we filled our tanks, there were scraping noises, followed by bad language. Investigating, we found the hose was leaking, a big patch of ice had formed and a succession of cyclists, coming round a corner, were losing control of their machines. It seemed unwise to reveal our part in the matter, at that time, so we took advantage of the darkness and confusion to remove our hose and retire to the boat, from which vantage point we commiserated with them on the vagaries of winter.

In more normal weather, the problems of living aboard tend to be mundane. Unpacking a locker to find out if something is in it can be a slow job, so keeping a record of what is in each one—though a chore, itself—is generally worthwhile. Stowing things so that they are not damaged, by each other or by water, takes care. Often we pack soft things and hard ones in the same locker, to prevent the hard ones from moving. But if they do, they may wear holes in the soft ones. So the sound of a jam jar, rolling to and fro, sends us looking for it, to wedge it securely before it has time to do any damage.

Clothing is easily damaged by chafe or water, so we do not take along more "good" things than we need. Usually we have one set of shore-going clothes for special occasions but otherwise we dress far more for comfort than for appearance.

Of course you need foul weather gear for storms but often, in ordinary weather, shirts and pants made of lightweight cloth are fine, with nothing over them, because any spray that lands on them dries out quickly in the sun and the breeze. Heavy sweaters made of untreated wool, which still has the natural oil in it from the sheep, are very good —if you do not let anyone clean them, which would take out the oil.

When living aboard a boat, getting laundry done is often a problem. In port, it means going ashore to a laundromat or, in a country where they do not exist, finding someone to do it. And at sea, you can either do it in a bucket, using a detergent that works in cold sea water, or tow it behind the boat.

Usually we tie a few items to a long line and tow them for one hundred miles, when they are declared clean by common consent. Then we rinse them in fresh water and hang them in the rigging to dry. But that procedure has its limitations—a shark may swallow the lot—so we only do a few things at one time. And if the line gets tangled with the log line, the spinner will wind the whole thing into a ball so tight, it is hardly worth retrieving.

Another factor is the need for constant maintenance. There seem to be endless jobs to do—in any boat—and if you fail to keep up with them, her condition soon deteriorates. So we always keep a list of the things to do next and each day, when we cross a few of them off, we have to add a few new ones.

Usually that is not a problem—it is a part of the job of running a boat—but on a long distance cruise, people sometimes try to earn money by working in ports along the way. Then it is common for the boat to deteriorate while they are working, until they need more money to get her back into shape, before they can continue with the cruise. And in some cases, the whole thing may fail because they never do get it straightened out.

So it is most unwise to start a cruise without having all of the funds on hand that you may reasonably need to finish it. For it is much safer to wait a whole year, if need be, and earn the money doing what you know, where you are known.

10

To the Great Lakes

The New York State Barge Canal provides easy access to the Great Lakes for yachts of any thinkable size (up to thirty-six-feet beam and twelve-feet draft) so long as they can get their overhead clearance down to fourteen feet for the passage. In a powerboat, that seldom means more than lowering a few antennae and taking down anything else that sticks up. But in a sailboat it means lifting out her masts and the best place to do that is City Island, outside New York.

The yards there know the procedure. You come under the crane and let go the rigging, while they set a sling just above the balance point of each mast. Then they lift them onto the dock, where you remove the spreaders and secure the rigging, while they make a wood horse to go on the foredeck, an X-frame to go aft, and a T-post to go in each mast step.

Those are lined up horizontally, about seven feet above the cockpit sole, and the masts are laid on them, with the heel of the mainmast forward and two thirds of its overhang aft. Then you secure everything, taking care to prevent the masts from moving fore and aft when the boat bounces in the wake of a tugboat, yet leaving the decks as clear as possible. And since you now only have one means of propulsion, you get an anchor ready to let go in case the engine should fail.

Then you need to buy charts for the Hudson River and the book of charts (put out by the Lake Survey people) for the Barge Canal system. After which you can proceed via Hell Gate and the Harlem River into the Hudson and head north for Albany. That takes two days but there is a good choice of places at which to stop along the way. And soon after you pass the city, you come to the Federal Lock at Troy.

As you approach it, you get lines ready fore and aft and put out a couple of fenders on each side of the boat. And standing well back, clear of any turbulence that may come your way when the operator lets the water out, you blow your horn three times to let him know that you want to go through. Then you wait until the gates open. After any boats that were in the lock came out, the traffic light turns green. And as you go slowly into the chamber, you find two men waiting to take your lines.

The lift in that lock depends on the state of the tide but it is never very great and the operator lets the water in slowly to give you an easy ride. So you turn off your engine to spare him the noise and fumes of your exhaust and tend your lines as the boat comes up. And when the gates open, you motor out into the calm, still waters of the Barge Canal.

Not far ahead is the point where the Champlain Canal goes straight on, while the Erie Canal that leads to the

Great Lakes goes off to the left. And just before that, on your right is the Troy Motor Boat and Canoe Club, where you put in for fuel, then go ashore and take a taxi to a feed store.

There you buy twelve large burlap bags, some line, and a bale of hay. Take them back to the dock, stuff the bags with the hay, and hang them all around your boat, on both sides, to protect her from the rough concrete walls of the locks ahead and save your fenders from being ruined.

In a powerboat, going through the locks is quite easy, since she is little affected by the motion of the water. But if her bows are flared, hang the hay bags high up to save the bows from the vertical walls and also low down, to protect her from the low walls that remain when the lock is full.

With a sailboat, your problems are different, for she is greatly affected by the surge of water as the lock fills. So you hang all the bags low down but use the fattest ones toward her ends, to discourage her from pivoting against the wall.

Just ahead are five locks, close together, with a total lift of over 150 feet. So there is bound to be a great deal of turbulence in the water, as you go up. And from now on, there will be no men to take your lines. But you could not control your boat in a lock that deep by lines hanging down from the top of the wall anyway. And the pegs in the walls, which barges use, are too far apart for you to reach. So you will have to use the ladders.

Each lock has four iron ladders let into the walls—one in each side, near each end. And the turbulence is always worse at the downstream end. So you use the ladder at the far end, on the side opposite the operator's little hut, where he can keep an eye on you as he lets the water in.

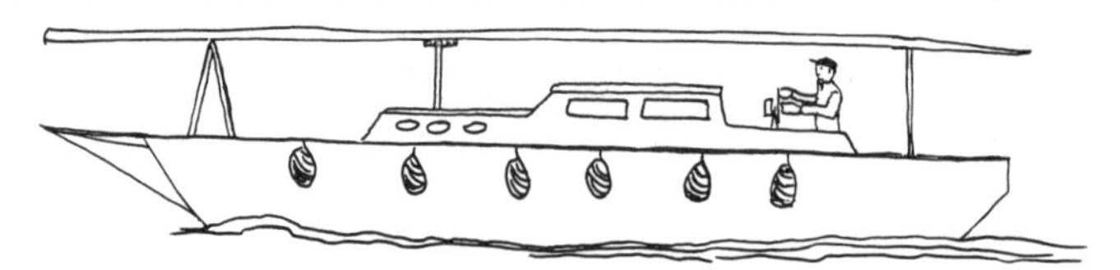

A sailboat with her masts and hay bags in place, ready to go through the canal from New York to the Great Lakes.

There you will secure the boat with a line that you move up as the water rises. But first you have to rig it. Take a docking line, make it fast at the bow, lead it aft outside everything, and make it fast at the stern, with four fathoms of slack in it. Then stand amidships and gather in the slack, until you have a bight two fathoms long that you lay on the deck. Do the same on the other side and you are all set.

Leaving the club, you motor across to the first lock and make the boat fast to the training wall, well back, so as to be clear of any traffic, while you walk up to the operator's house and get your permit to go through the canal. There is no charge and it only takes a minute but you will be asked for its number at various points along the way.

Back aboard, you wait for the green light, go very slowly down the lock, and stop alongside the ladder. There you tuck the bight of your line up behind a rung (about eye level) and pull the slack over it. Then turn the bight over to cross the lines and open it up so that two people can stand about eighteen feet apart, facing the wall and holding it, as shown in the drawing.

That gives you good control of the boat, for the hay bags take care of her pressing against the wall, while you can keep her from pivoting (which is the worst problem) by pulling on the line or pushing on the wall—with a hand, if she takes a little sheer or with both feet, if she takes a good one. But never get a hand or foot between the boat and the wall.

As the water rises, you wait for a calm moment, then go to the ladder, quickly move the bight up a few rungs, and get back to your positions before anything happens. But as the lock fills, the turbulence dies down, until you are bobbing gently alongside the low wall at the top.

At the next lock, you again see which side the operator's little hut is and head for the opposite ladder (which is why you have lines on both sides) and by the time you get through the first five, you should be quite good at it. Then you come to a long level and can relax for a while.

On the charts of the canal, you will see terminals where the barges used to unload. But now they are seldom used, so they make fine stopping places for yachts. Each one has a long wall, usually with a grassy area behind it, but the bollards on it are too far apart to use in the ordinary way. So you bring your stern level with one bollard, run a breast line to it and a bow spring back to it, then take a line out

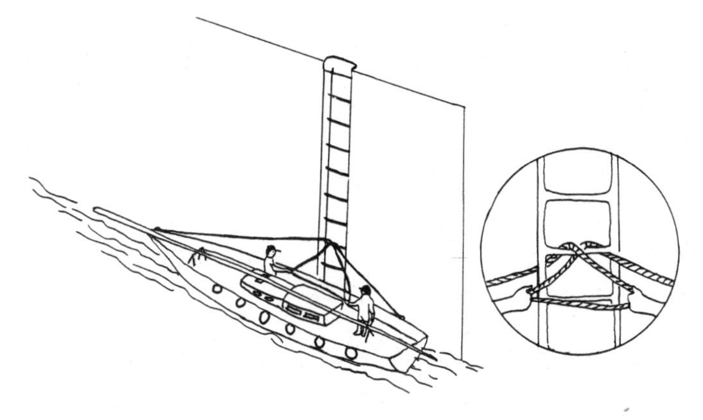

The boat is in a lock, secured to a ladder by a line held by two people—standing well apart—as described.

ahead to the next one.

The bow line and the spring keep the boat's bow in and prevent her from surging, while the breast line holds her stern in, and the hay bags (that you leave in place all the way) keep her off the wall. And when an oil barge goes past in the night, pushing the water ahead of her, it does not bother you.

Some of the terminals have been made into marinas, with fuel and water and other facilities available, so that running the canal presents no problems. But the speed limit of ten miles an hour is enforced and when planning a day's run, you need to allow twenty minutes for each lock you have to go through.

Few of the locks are as high as those behind you, except the one at Little Falls and that is not as bad as it looks, for the walls are smooth and the operator lets the water in gently. And locking down to a lower level is a breeze.

You still use the far ladder, because there is a sill at the upstream end, but since the lock is full when you go in, it is easy to come alongside the wall and step off with your line. And the water goes down smoothly, so that all you have to do is keep moving the bight down the ladder, checking it now and then to stop the boat from drifting away from the wall.

Beyond Lake Oneida, you come to Three River Point where the canal divides, the right-hand branch leading down to Oswego on Lake Ontario, while the left-hand one continues on to Buffalo on Lake Erie, via Tonawanda Creek and the Niagara River.

The current in the Niagara is no problem until you come to the narrow part and there is a canal along the south bank to take you past that, into the lake. But with a sailboat, you stop at a yard just before the canal and put the masts back in. This time, they lift the masts onto the dock and let you prepare them, before they step them. And as soon as the rigging is hooked up, they leave you to do the rest.

In the canal there is a lock and just past that, on your left, is a building where you can buy all the Great Lakes charts. But the single-digit one for each lake, plus the two-digit ones of its coastline are enough, since the latter have inserts that give you details of the principal harbors. And farther up the canal is the Buffalo Yacht Club, where you leave your hay bags, before going out onto the lake.

There is a good weather service in the Great Lakes, which you can get by calling the nearest Coast Guard station. The day is divided into six-hour periods and the information is given in a code which is available at any yard or marina. Each five-digit group gives you the anticipated wind direction, its force, and the visibility for a designated area, for one period, and they can usually give you them for two periods.

So in a powerboat, you can make day runs from port to port, starting right after the 6:00 A.M. forecast, with a fair degree of certainty that you will not get into trouble. But in a sailboat you have to watch the weather more closely, as you would on a coastwise passage at sea.

In fact, the lakes are so large that sailing them is much like being at sea, with a few notable exceptions.

The charts are different. They have courses laid out on them, in statute miles that you have to multiply by seven and divide by eight to convert to nautical miles. And cribs, which turn out to be square concrete structures, sticking a few feet out of the water, for various commercial purposes. But there are few shoals or off-lying dangers and navigation is a great deal easier.

There are virtually no currents, except in the narrow places between one lake and the next, where they flow steadily in one direction. And no tides. But the lake levels vary from year to year. (Check that at the chart place in Buffalo.)

The navigational aids are good, though the first time we sailed Lake Erie, we were confused by the Perry Monument. In real life, it looks like a tall lighthouse; on the chart you can hardly find it. And Port Huron has a black lightship, which you could easily mistake for a small freighter.

There is a phenomenon called lake steam that looks like dense white fog but lies low on the water, so that if you climb up a few feet, you may see a ship going by in the sunshine above it. Or green trees standing in the mist along the shore.

And the shipping lanes are so busy that on a clear night you can often see the lights of a dozen freighters at once.

There are no locks between Lake Erie and Lake Huron, or between Lake Huron and Lake Michigan, so that from Buffalo you have clear sailing, all the way to Chicago. But to get to Lake Ontario, you have to pass through the Welland Ship Canal and if you are coming from New York, it is easier to turn off the main canal at Three River Point and go down to Oswego.

From there you can cross the lake and go down the Saint Lawrence River, through the Thousand Islands, to Montreal. The locks in the Saint Lawrence are meant for ships and you have to wait to go through each one. But they do have floats let into the walls—for small craft to more to—which go down with the water and make locking through very easy.

At Sorel, just beyond Montreal, you can leave the river

and take the Richelieu Canal up to Lake Champlain (if your boat is less than twenty feet wide). From there you can go down the Barge Canal to Troy and follow the Hudson River to New York.

But the Great Lakes are for the summer. The winters there are long and cold. And though the Barge Canal system may be open for nearly six months in a mild year (you can get the dates from the United States Army, Corps of Engineers, in Albany) it is best to take a month off each end, when planning a trip there.

11

Collision Courses

Whenever another vessel is coming closer to you and her bearing is not changing, you are on a collision course.

Usually it is obvious and you are soon aware of it. For she seems to stay in one place, while everything else changes. And each time you look at her, you see more details.

So if she is another yacht, you either give way to her or hold your course and speed, as required by the Rules of the Road. But if she is a ship, it is different. For she can not stop, or even change course, quickly. She has too much inertia.

This means you have to get out of her way. But you must do it in good time and make your intentions quite clear.

She is unlikely to hear your horn or answer your radio but in coastal waters she can probably see you. So you make an exaggerated turn, hold that heading for a moment, and straighten out on your new course. If she does not like it, she will tell you so with her whistle. Otherwise you continue on your way but keep an eye on her until you are out of danger.

In restricted waters, you know exactly where a ship will go, because she has no choice. She has to stay in the narrow channel where there is enough water to float her. Yet in the Detroit River, that runs between Lake Erie and Lake Huron, we saw dozens of small boats going to and fro across the bows of fast moving ships. So we asked if they were ever run down and the answer was: "Yes, all the time."

Which clears that up. If you cross such a channel at the wrong time, you may be killed. You can get the same effect with highways and railroads. But outside the channel you are safe, because no ship can reach you. And in most places, you can go along one side—just beyond the buoys —in plenty of water, if you follow the charts carefully.

But watch out for the unlit buoys that are sometimes put near the lit ones as "markers" in case they drag out of position, for those are seldom shown on the charts.

Offshore, it is best to assume that a ship has not seen you, unless you are quite sure that she has, because a yacht is hard to see from a ship's bridge, even if her crew were looking for one and why should they be doing that?

With radar, they can "see" another ship long before she comes into view, so they watch the screen. When a "blip" gets close, they will look out of the window and see what it is. Otherwise the view is dull and even a man staring straight at you may not see you, if his mind is somewhere else.

Unfortunately a yacht's tiny echo is easily lost in the ground clutter on a ship's radar screen if there is any kind of a sea running. And although a reflector helps, it would be most unwise to assume she will see you in time to avoid you.

Again, it is best to get out of her way. But to do that, *you* have to see *her.* And a fast ship can come out of nowhere and be on top of you surprisingly quickly.

So the helmsman should stand up every fifteen minutes and look all around the horizon. In calm weather he can do it in one continuous search but in rough weather he has to take a sector at a time—as the boat comes to the top of a wave and he can see out across the water—until he has covered a whole 360 degrees.

When he finds a ship, he takes her bearing, by sighting over the main compass. And a few minute later, he will do it again. If the bearing has changed, there is no problem, though he will keep checking it, from time to time. But if not, he must do something about it, right away.

For a twenty-knot ship five miles away will be where you are in fifteen minutes. And at six knots, you can go a mile and a half in that time. So you want to sail at right angles to her course, in the direction that will take you farthest from her track.

To determine that, you first look at the ship. If she is heading slightly to one side of you, even though her bearing is not yet changing appreciably, it will almost certainly pay you to go the other way. For even five degrees will put her nearly half a mile to that side, by the time she reaches you. But if she is coming straight toward you, head whichever way you can go fastest, depending on the wind and sea.

Often it seems that the ship is turning the way you are going, because the angle at which you see her changes so slowly at first. But if you take her bearing every few minutes, you can figure how the situation is developing. Usually it gets better all the time. But once in a great while, a ship really will follow you, having decided you need rescuing.

That is embarrassing, for she comes up on your windward side, meaning to shelter you but in fact leaving you rolling and slatting in her lee. And if your engine will not start, you may well be in danger as she drifts down onto you.

Meanwhile she is hailing you but she will never hear any reply you make. And if you wave her away, she may think you are asking for help. So you go through a little charade.

Calling the rest of your crew on deck, you sit around in the cockpit, pretending to drink coffee out of empty cups, yawning elaborately, and doing anything else you can to convince her that you are all right, until she goes away.

Fortunately that never seems to happen at night, perhaps because the ship does not see you. For the running lights on a sailboat are of little use, except to other small craft. And if a ship is going to miss you, there is no point in complicating things by attracting her attention.

But in coastal waters, when there are several ships in sight at one time, we turn on our spreader lights now and then, or shine a lantern on our sails, to show everyone what kind of vessel we are and which way we are going.

Ships are easy to see at night and their range lights make it clear where they are headed. But the smaller vessels can be confusing, espcially tugboats and fishing boats.

In the distance, the lights of a fishing fleet look like a village but closer, you can see they are moving and when you get there, you find each boat going to and fro, while her crew work their nets under bright overhead lights. So you weave your way through them, judging their courses by their movement and trying to guess what each one will do next.

A tugboat normally holds a steady course but the lights on her tow are often barely visible, fifty feet away. So it is best to stay well clear of her. And if she has a long tow, be very careful about crossing her stern.

One misty day in the English Channel we saw a tugboat go across our bows but no barge followed her. And when we came to her wake, there was no cable to be seen. Then suddenly it lifted out of the water and rose into the air, leading up to the bow of a large ship that was coming silently toward us.

Another thing to remember is that when small commercial vessels meet nearly head on, they often pass starboard to starboard if it is more direct than going port to port. But when in doubt, take bearings and see how they change. Even if the other "vessel" is a lighthouse or a waterspout.

There is a narrow, twisting channel between the island of Ushant and the coast of France, where the tide runs so strongly that the lighthouses leave wakes like ships. And going through it on a flat calm day in a sailboat with no engine, we used the wind created by our own motion to dodge them. But the principle was the same: we took the bearing of each lighthouse as it came rushing toward us and treated it like an oncoming ship. Then as we went swirling past it, we started taking bearings on the next one. And so on, down the line.

Off the coast of Cuba, we saw several waterspouts coming toward us, borne along on the wind and moving quite fast. So we used the same technique, taking bearings on each one and seeing how they changed, then laying our course to avoid them. And it worked well, carrying us safely through them.

Still there are special cases, like "Tanker Alley" where the northbound ships ride the Gulf Stream past Florida and you may see four of them, in line abreast, bearing down on you. There it is best to treat the axis of the stream (which is on the chart) as if it were a ship channel. As you approach it, you look for a gap in the traffic, then cross it as fast as you can. Just like the highway it is.

12

On Pilotage

Pilotage is the art of directing a boat by visual contact with known objects on the land or afloat. It falls naturally into two parts. The first is finding out where you are. The second is figuring how to get where you want to be, without running into anything (like a rock) along the way.

So you need a chart that shows where the objects are and a means of finding your position, relative to them.

Except at very short ranges, it is not practical to measure your distance from anything by eye. So you have to rely on bearings, taken either with the boat's main compass or with a portable one, made for the purpose.

The portable ones usually have prismatic sights and can be aimed quite accurately, but they are subject to errors because of anomalies in the magnetic field around the boat. So before you use one, you have to find a place where its readings agree with those of the main compass (which you have checked) and thereafter you must always use it in exactly that place. But you can not use it near the main compass, because they fight. And in most boats, it is hard to find another place from which you can take bearings in all directions. So more often than not, it is better to use the main compass.

For that you take the cover off the binnacle and stand where you can see the object over the compass. Glance down and move your head a little, until you are looking exactly across the center of the card. Then look up again at the object and down again at the card a couple of times and read the bearing on the far side of the card, in line with the center.

In fact, the whole thing only takes a few seconds, so you note the reading and take some more shots. After a little practice, each reading should be within three degrees of the proper bearing and the average of four shots will be accurate to one degree, which is good enough.

The main compass has the further advantage that it is always available (you do not have to go and get it) and since it was checked, it has a correction card. But when using the card, remember to apply the correction for the boat's heading at the time, NOT for the bearing you were taking.

In places where the marks are near by, it is sometimes possible to take bearings of acceptable accuracy by judging the angle between the object and the boat's bow and adding that to (or subtracting it from) her heading. For while an error of one degree will put you out by a tenth of a mile when the object is six miles off, it takes an error of nearly six degrees to cause the same effect when the mark is only a mile away.

But it is always important to be aware of the degree of accuracy to which you are working. Otherwise you may be taking risks that you do not know you are taking.

We never regard a fix made of two elements (two bearings, or a bearing and a depth, etc.) as being definitive. But a three-element fix, in which everything comes neatly together, can usually be trusted, as can a succession of two-element ones that agree with the courses and distances between them.

The first rule with charts is to have enough of them. If an emergency arises and you have to put into the nearest port, there may be no time to feel your way in. And while a problem that arises far offshore must be dealt with there, it would be embarrassing near shore not to be able to take a sick man to a doctor, just because you did not have the proper chart.

So for a typical passage, we carry the approach charts to one or more intermediate ports, as well as those to our ports of departure (we might want to go back) and arrival. And of course the sea charts, on which to plot our position en route.

And since we find it impractical to keep them up to date, we never use a chart more than two years old.

Foreign charts can be confusing at first, since they are usually printed in black only, on white paper and the buoys are hard to find among the soundings. And when you get there, the black buoy on the chart may well turn out to be rust red at the bottom and seagull-manure white on top. But they do give you one break: they often include views of harbors and so on, in their margins, as seen from the deck of a ship at sea.

Those are a great help, for it is not easy to convert a plan view of a harbor, as given by the chart, into the kind of vertical view that you actually see as you approach it.

But it is a most useful skill to have, since it lets you know what you are looking for before you see it. So it is

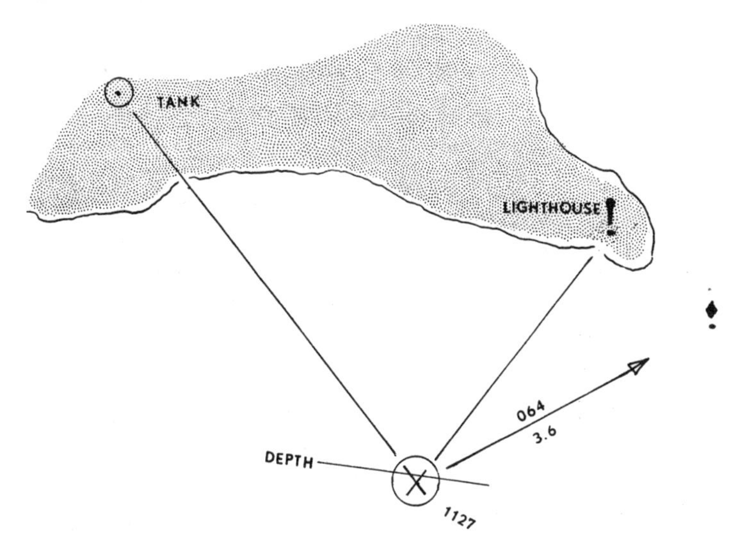

A three-element fix like this—made of two bearings and a depth —usually results in a small triangle, the center of which can be taken as a point of departure for the next leg.

well worthwhile to take every opportunity of comparing the chart you are using with what you see, until you reach the point where you can pick up any chart and immediately "see" at a glance what the coastline will look like from any position.

Even with the best charts, we are cautious about fixing our position, for it is so easy to goof. And the easiest way of all is by taking a mark, assuming it is the right one, and ignoring any others that may be in sight.

So whether we are taking bearings of distant steeples or heading toward a buoy, we always double check. Is that really the steeple we think it is? Could it possibly be any other? If it could, it probably is. And that buoy. Have we accounted for every buoy in sight? Do the bearings of each one agree with our assumed position? If not, we may be all wet.

At night, we always confirm the identity of lights. An easy way to do that is by counting caterpillars.

If you say: "One caterpillar, two caterpillars, three caterpillars" and so on, about as fast as you conveniently can, you will be counting seconds. (Check it against your watch, to make sure you have the right-size caterpillars.) So without shining lights or otherwise upsetting your night vision, you can tell what the code of a light is. And if there is any doubt, you can always use a stopwatch to get a more accurate figure.

But still you have to double check. One time, off the coast of Spain, we picked up a light that seemed to flash twice every thirty seconds. But when we checked it, it only flashed once. And there were no thirty-second lights anywhere near where we were. But there was a sixty-second light. So we looked it up in the Light List and found the answer: It flashed once and thirty seconds later it flashed twice. And thirty seconds later it flashed once. And so on. But on the chart it said: Fl (3) ev 60 sec.

Finding your position by night is often much easier than it is by day, because the lights reach out so far. And as you approach it, a single lighthouse can give you a fix. All you have to do is look up its height, allow for your own height of eye, and look in the tables to find how far away it will be dipping. Then you pick up its loom, wait for it to start flashing (intermittently, as it dips below the horizon), and take its bearing. And you have a fix that is quite good enough, in most cases, to take you to your next mark.

But in close waters, like Chesapeake Bay, there may be so many lights of various kinds in sight at one time that the problem is to sort them out. And in that case, taking the bearing of each one is the easiest way to do it.

Remember that aids to navigation are just that. Nothing more. They do not do the navigation (or the pilotage) for you. They are put there to help you do it. And if you go blindly from one to the next, you will sooner or later run aground.

For it is not practical to install and maintain enough buoys and beacons to guard every shoal in a harbor, every bend in a river. So the funds available are used to place marks where a competent navigator will find them most useful.

In fact, you can never be sure that a light will be lit or that a buoy will even be there. Entering a small harbor in France one day, we could not find a single one of the buoys that were supposed to mark the narrow, rock-bound channel. And when we got there, we found them all on the quay, being painted.

So how did we get in? We used the things that never move, that man can not change. We took bearings on headlands and large, prominent rocks that were clearly and unmistakably marked on the charts. And we knew exactly where we were, all the way.

When you know where you are, it is easy to figure how to get where you want to be, after studying the chart. But the best route to use will often depend on other factors, like the state of the tide, the wind, and the visibility.

So think about it. Compare the virtues (and evils) of all possible routes and then decide which one to take.

When that is established, you can direct the boat in much the same way as you would by any other type of navigation. First you lay off your courses, then look up the currents, then figure the heading, distance, and running time for each leg.

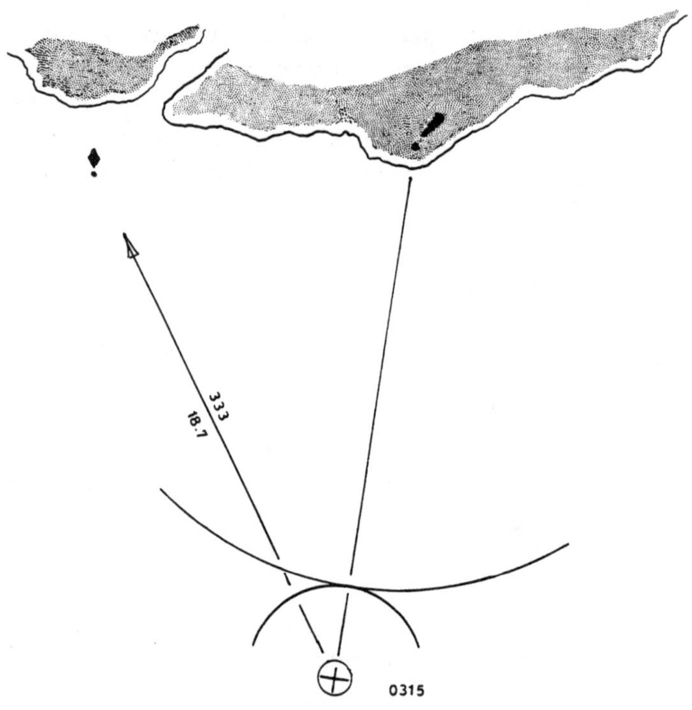

A one-light fix: When a light is dipping, its distance is that given in the tables, corrected for your height of eye, and the fix you get is usually good enough to take you to a safe landfall.

Just because you can see where you are, it does not mean that you can afford to be sloppy. Always calculate; never leave anything to chance. And if there is a discrepancy between your calculations and the performance of the boat, find out why. Perhaps the current is stronger than you thought? Or maybe weed on her hull is slowing the boat down? You have to know.

In open water, when you know for sure what the current is and how fast the boat is going, running fixes are useful. In the simplest case, you take the time when a mark is four points off the bow and again when it is exactly abeam. And the distance you ran between those times is your distance off the mark.

But you can also get a running fix by taking the bearing of a mark at any time, treating it as a position line, and advancing it to cross another, from the same mark, later. And if you have a depth finder, the bearing of a mark plus the depth of water will often give you an immediate fix.

In restricted waters, you have to calculate more quickly but since the distances are shorter, a lesser degree of accuracy is acceptable, so we use our own shorthand navigation. For that we follow the same proceedures but we abbreviate everything. To measure distances on a chart, we use two fingers as dividers. Courses we estimate by eye from the nearest line of latitude or longitude (with variation and deviation added). And all calculations are done in one's head. But since we are doing the same things as always, we are unlikely to make mistakes and if we do, they look wrong.

Near the shore, you can often take advantage of the tidal currents by moving into deeper water when they are going your way and standing onto the flats while they are running against you. And it usually pays to set your time of departure so as to carry a fair tide where it runs strongly.

But remember that currents also run on and off the shore, as the tides fill and empty large enclosed areas of water, so that you may be carried sideways as you go past them. And beware of the very strong currents that run on and off large, shallow banks, such as you find in the Bahamas.

Another problem in areas like the Bahamas is that the

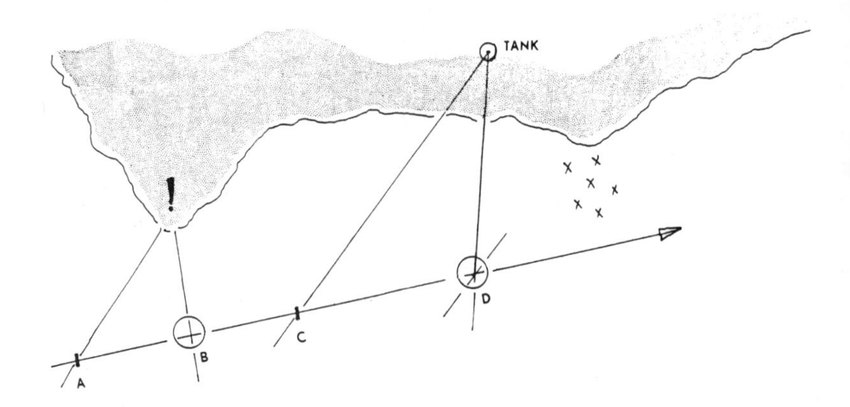

Two kinds of running fix: The distance from A (when a landmark is four points off the bow) to B (when it is abeam) is the same as that from the landmark to B. But you can also get a running fix by treating a bearing from C as a position line and advancing it to get a fix at D in time to avoid hitting the rocks ahead.

charts are not always accurate. In fact, you may find an island as much as twelve miles out of position, by celestial navigation. And though the weather is usually fine, the visibility may be no more than three or four miles because of haze. But in the Windward and Leeward Islands, the visibility is generally so good and the mountains are so high that by day you can often see from one island to the next.

Navigating by radar, you are in visual contact with everything it can see. And though individual landmarks may be hard to identify, it is usually quite easy to match the general shape of the coastline with that shown on the chart. But again you should always measure and calculate, then check the results against any other clues you can get, like depths of water or brief glimpses of the land. For anything that complex can easily go wrong and sometimes when it does, it gives no sign that it has.

In small waters, where there are cross currents, it pays to make your own ranges. Almost anything will do, like a tree in line with the side of a house, so long as one is well behind the other and they take you where you want to go. For at short distances, ranges are very accurate indeed and if you should be set sideways by even a few feet, you will know it.

If you are approaching a harbor and there is only one mark you can be sure of, look on the chart for a bearing from it that goes out to sea, clear of all dangers. Then as you go in, keep the mark on the reciprocal bearing and you are safe.

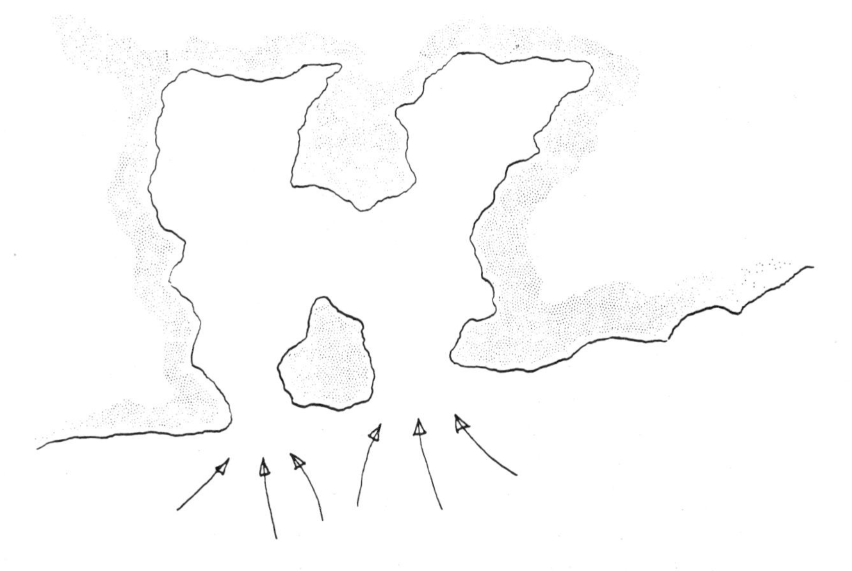

Strong onshore and offshore currents can be expected where the tides fill and empty large enclosed areas of water.

And as you leave a harbor, always note your course and heading. Then if you run into a problem (like a wall of fog), you can make a 180-degree turn, apply your drift correction the other way, and pick up your marks as you go back.

When there is only one landmark you can be sure of, it is best to approach the harbor on the reciprocal of a safe bearing from it—one that takes you clear of all dangers.

13

Paper Work

Charts often come rolled up, which makes them impractical to use, so we fold them to convenient sizes and stow them under the seat cushions in the saloon, where the people's weight soon flattens them out. It is good to have plenty of them, so that a sudden change of plans—as might happen in an emergency—does not leave you unprepared. And you also need books.

The light lists, pilot books, and tide and current tables for each area and time span are essential. But they are mostly government publications, intended for larger vessels, so what they say may need translating, for use in small ones.

For example, a pilot book may recommend a harbor so large that when you get there, you can barely see the far shore—and have little shelter from the weather—then go on to say that a creek nearby is almost impossible to get into. But when you try it, you may find a good harbor, full of yachts. So it is a good idea to read what a pilot book says about parts of the coast that you already know. That way you can get an idea of how the writer thinks before you use it. Of course, the author of a foreign pilot book may take an entirely different view, so you have to be cautious at first. And even within the United States, the siting of the navigational aids varies from place to place.

On the intracoastal waterway there are sudden changes in the philosophy behind the location of the aids. In one district they are on the outsides of curves, attracting you toward them. In another district, they are on the insides of curves, warning you off the shoals. But once you understand the mind of the man who put them there, it is easy enough to use them.

The books of tidal and other information that are printed commercially for small craft are often handy. And tidal current charts—usually printed in book form—are very useful. But if no information is available on the current at any location, you can generally figure it out from the tide tables.

For chart work, we have a ditty bag, containing dividers, pencils, erasers, a sharpener, a pair of compasses, a protractor, and sundry other items, collected over the years. It always goes with us on a passage and if anything we need is missing from the navigator's "office" aboard a strange vessel, there is always what we need (or something that will do) in the ditty bag.

The ability to make do—to jury rig almost anything—is most valuable in a cruising yacht. Things will break and you can not carry two of everything you may ever need. So it pays to keep in mind the basic purpose of anything, and how that may be achieved in quite a different way, if the need should arise.

No procedure or piece of equipment is sacred. To treat

a sextant, for example, with reverence is as silly as worshiping a graven image—and about as effective. It is a useful instrument that should not be dropped but if it is, you can find what error it has and usually get it working well enough to take you to the next port, where it can be repaired and adjusted.

And surely it is even sillier to pretend that conventions or nomenclature are anything more than parochial conveniences. A cruising yacht seldom has more than one navigator and whether he writes her heading above, below, or beside a line on his chart is of not the slightest consequence to anyone else.

Since the crews of most yachts spend far more time ashore than at sea, they are unlikely to react as quickly to a nautical term as to a more familiar one, so it is logical not to use more nautical terms than absolutely necessary. And sometimes we use a term of our own, if the circumstances require it.

On the intracoastal waterway, for example, there are many daymarks and lighted structures, some of which look alike in the distance. So we call the daymarks "markers" and the lighted ones "beacons"—two quite distinct words—which enables us to refer to them quickly with no chance of misunderstanding.

And if someone criticizes your use of words, you can give them the line Winston Churchill used, when someone told him that he could not end a sentence with a preposition: "That is a piece of pedantry, up with which I will not put."

14

Dealing with Sharks

There are roughly 250 different species of sharks in the sea but 85 percent of them are either too small, or too sluggish, or do not have the right kind of teeth to eat people.

That leaves about thirty-seven species, of which twenty-seven are known to attack swimmers and the rest could do so if they felt like it. In fact, some of those may well be killers, for the records on the subject are far from complete.

In an average year, about forty attacks by sharks on people and six more on boats are recorded by various agencies. But there is no requirement to make such reports and in many cases, there may be no witness to give it. For it is known that sharks prefer to attack a single swimmer, rather than one of a group swimming together. And if a lone fisherman's boat is attacked, who can be sure that he was lost to a shark?

So it is quite possible that there are many more attacks than show up in the reports. But still the numbers are small and while the thought of being eaten alive is less than pleasant, the chances of it happening are in most cases so slight as to be worth taking, rather than give up something else.

At one time, I used to swim alone off a remote beach in Barbados every day, after work. I would go out with a face mask and a snorkel and wander around, admiring the brightly colored fishes that darted in and out of the coral. And in due course, my friend the shark would come by.

Sharks look bigger under water but still I would say that he was much larger than me. And I never doubted that he could eat me, if he wished. But he did not seem to want to.

He would come down the beach from the north and go past me on the seaward side, maybe twenty feet away, cruising steadily along with easy, graceful movements. And though he never changed his course or speed, his eye (on my side) would look straight at me, turning to follow me as he went by.

The first time I saw him, it was a shock. For my spear gun would be useless against him (except to prod him in the nose with, as a last resort). But if you swim alone in the Caribbean, you are liable to meet a shark. And it is possible that he will attack you. That is the price of admission.

But it was hot ashore. And the beach was quite beautiful. And beneath the surface of the water were so many things I had not seen. So of course I went back the next day. And in three or four months, I got used to him.

On the other hand, there is no point in taking any more

risk than is necessary to do what you want. So if you are going to swim where sharks may be, you should know more about them.

From a shark's point of view, the surface of the water is like a silvery, undulating mirror, with a round patch directly overhead, through which he can see the sky above. So a man with his head out of the water is an interesting item of food, dangling through the roof. But a man with his head under the water is another being in the shark's world—a strange one, that it might not pay to fool with.

Which would you rather be?

Like most fishes, a shark will commute, if he can. In the morning, he will go where the food is and in the evening, he will return to his resting place. But unlike the others, he has no swim bladder to keep him afloat. When he stops swimming, he sinks. And the only way he can get any rest is to find a place where he can lie comfortably on the bottom.

At one time, it was common practice for fishermen to find a sleeping shark, swim down, and put a noose around his tail. But that proceedure has fallen into disfavor of late.

So long as there is plenty of food, a shark is likely to continue his commuting routine indefinitely, for his intelligence is quite limited and his interests hardly go beyond food, sleep, sex, and curiousity about things that shine or wiggle. But he needs a lot of food and if he does not get it, he becomes desperate. Then he will roam far out into the ocean, going for long periods without sleep, and becoming more irritable all the time. And then it does not pay to swim with him.

So while most sharks are seen inshore, where the food is plentiful, relatively few of them attack swimmers. But offshore, where food is often scarce, they can be highly agressive.

One day, when we were becalmed in the Sargasso Sea aboard *Wind Song* (a thirty-eight-foot ketch out of St Thomas for New York), a shark repeatedly attacked the boat. He would swim slowly away, maybe fifty yards, then turn and come toward us, gathering speed to ram us as hard as he could. And when he hit, the whole boat shook.

It can hardly have been pleasant for him, hurling himself against a large, heavy boat and I believe he only did it because he was desperate. He was determined to break open the boat and get at the food inside it. Which happened to be us.

Wind Song was an elderly, wooden boat and she had been leaking quite badly before we were becalmed, so I lay in wait for him with a long, heavy boathook that had a sharp, pointed end. And as he came in, I hit him as hard as I could on the nose. The boathook jarred and bounced back as though it had struck rock but he left immediately and never came back.

Of course, a shark could become desperate from hunger in other places, for other reasons. Perhaps there is a local shortage of food. Or too many sharks to share it. Or he just did not get his share. Kingston, Jamaica is a case in point. There the entrance to the big harbor is narrow and tricky, so that sharks who follow ships in, often can not find their way out again. And over the years, a shark population has built up that the available food can not sustain. So most of them are desperately hungry, most of the time.

This indicates that you should seek local advice, whenever possible, before swimming where sharks might be.

It has been said that sharks are totally unpredictable,

but when you take into account their mental level (stupid) and their motivations (hunger and curiosity), their behaviour makes sense. In fact, the surprising thing is that they do not eat more people, since we know they like pork.

One possibility is that a shark is more timid than we realize (unless his fear is overridden by hunger, or forgotten in a moment of curiosity). For his kind have been around for millions of years and deep in his subconscious there may well be memories of other, more powerful creatures, now extinct, from which his ancestors had the good sense to flee.

Another possibility is that a human being, in the water, reminds a shark of an octopus, which is his natural enemy. Approaching the coast of Ceylon in a small open boat, we were accompanied for some time by a very large shark who circled us constantly. Then suddenly he went rushing straight out to sea at high speed. Later, ashore, we asked why that was and we were told that he must have seen one of the giant octopi that live in the rocks there, underneath our boat.

But in any case, a shark's behaviour will depend to a large extent on the species to which he belongs. And the most dangerous are the white shark, the hammerhead shark, the tiger shark, the blue shark, and the sand shark. From the records of attacks on people, it would appear that those species all prefer water temperatures of seventy degrees or more, which would confine them to the tropics except in the summer. But the white shark, which is the most dangerous of all, seems to be at home in far cooler water. So while swimming in cooler water is safer than it is in warm water, there is nowhere in the sea that you are completely safe from sharks.

Any shark that attacks man may start doing so when

he is quite young, say three or four feet long. But the largest species, the basking shark and the whale shark (which grow to forty and sixty feet, respectively) are not dangerous to swimmers. However, a basking shark likes to bask on the surface of the sea; if you run into one with your boat, he will attack the boat.

You are quite safe from sharks aboard a large, heavy boat. Until you fall overboard. Or decide to go for a swim. Or have to clear a line from a propeller. But in a small, light boat you have no such immunity. So before crossing the Atlantic in *Sopranino,* we fitted a guard rail around her cockpit and in tropical waters we never fished. For if you cut up a fish and throw the blood overboard, it will very soon attract evey shark for miles around. And getting rid of them may not be easy.

We were given some black powder to throw in the sea, that imitates the ink which an octopus puts out. But we were told to be careful about using it, since it might attract octopi. Ernest Hemingway kept a box of hand grenades aboard *Pilar,* which he said were effective. But shooting at sharks with a rifle is a waste of time, because refraction makes it hard to aim and the bullet slows down very quickly in the water.

In fact, as long as no shark tries to break into the boat, we prefer to do nothing, keep a low profile, and wait for them to go away. But if one did attack, I would hit him on the nose with a boathook, as I did on *Wind Song.* For a shark's nose is his most sensitive part and you are unlikely to miss at that range.

And though we only noticed sharks around in calm weather, they ate our log spinners all the time. We painted the spinners black, to make them less attractive but still the sharks ate so many, each year, that we took them off our income tax.

If you decide to swim where sharks may be, it is safer not to do it alone. When swimming from a boat, you should have someone on the boat as a lookout.

It is most important to keep your head under water, so you need a snorkel to breathe through and a face mask to help you see properly. We also use swim fins, which give you much more control in the water. If we are not working with our hands, we carry something at least three feet long that we could prod a shark on the nose with, if we had to. (A knife is very romantic but we would rather not get that close.)

In a dim light, sharks can see better than people, so it is safer to swim in full daylight. But never wear bright colors or carry anything shiny, like an exposed knife. And in the water, do not make any sudden, jerky movements that would attract the attention of a passing shark.

With a face mask, your vision is limited, so that you have to keep turning your head to see what is around you. If you see a shark approaching, try not to panic. Remember, he is probably just curious. After all, you do look odd to him. So he is trying to figure you out. And you must convince him that you are neither hostile nor afraid.

So turn and face him squarely. No sudden movements, just ease around. Bring your weapon up in front of you—between you and him—but do not threaten him with it. Then keep still, look him in the eye, and hope that he will leave.

Usually he does. But a couple of times, I have met ones that did not and in each case I backed away slowly, moving my fins gently up and down—the shark following but not coming any closer, still hesitant—until I reached safety.

Safety?

The first time, it was a beach. As I moved into shallow water, the shark turned and left. And the second time, it was our boat. I saw the anchor rode in the water and using it as a reference, I maneuvered into position by the ladder. Then I got ready and quickly went up it, into the world of man.

15

Rope Tricks

Aboard a yacht, you need to know only nine knots and most of them are related to two basic ones (the *bowline* and the *half hitch*) so it is better to learn those well—until you can make them with your eyes shut—than to spend time on a wide variety of knots, which you may forget unless you use them.

If you make a *bowline,* then toss the bight over something and pull on the line, you can see how it works: the harder you pull, the more firmly one part of the line grips the other. Yet you can easily undo the knot, by pushing it apart.

A *common bend* (or *sheet bend*) is the same knot, made with two separate lines, instead of one, and is a good way to join the lines, even if they are of different sizes. So you have two excellent knots to use when you can make a bowline.

The *half hitch*—easiest of all—works in another way and needs a second one to lock it. *Two half hitches,* as they are commonly used, are the same as a *clove hitch.* All you need do is make them around a spar or toss them over a bollard.

The *fisherman's bend*—for attaching an anchor to a rode—has two round turns to take the strain at the shackle, but then it is just a half hitch through the turns and another one beyond them, to be finished off with a seizing for safety.

A *midshipman's hitch* is like two half hitches but with an extra turn, through the first one. Once made, you can slide the knot along the line (making the bight any size you want) and it will stay there. So it is useful for tying things down, because you can easily take up any slack that develops later.

A *rolling hitch* is the same as a midshipman's one but made around something—like a spar or a pile—and is useful because it will not slip lengthwise. You can use it to lift a mast, with a crane, or to attach the end of a dinghy's painter to a dock at a point high enough to reach when the tide comes in.

The *figure eight*—another easy knot—is very useful for preventing a line from going through a block or a deadeye, so we always put one in the end of any sheet we are using.

But the *reef knot* (or *square knot*) is easy to spill, so it should only be used where that is an asset, like the reef points in a sail, and never to secure anything important.

There is little need for splicing in a modern yacht but it is useful to be able to make an eye splice, in an anchor rode or docking line (that may be laid or woven). And the same splice is sometimes handy for making jury rigs, in emergencies.

One old fashioned thing we still use, aboard larger boats, is a *handy billy.* That is a small tackle (just two blocks, with a hook on one and a tail on the other) you carry around and put on anything you want to move, to get more purchase.

We usually make them a couple of fathoms long (when fully extended) and secure one end of the line to each block, so that we will not have to look for it, in an emergency.

When you have to whip the end of a line or seize two lines together, the Irish method is good: You start at the middle of a piece of twine, wrap it around the line (or lines), and make half a reef knot. Then you take the ends around to the other side and make another half reef knot, then go back to the first side, and so on, until the whipping or seizing is long enough. It may look rather crude but it holds better than a conventional one. And it can not fall apart quickly.

But when you make a lashing—around things that are some distance apart—it is better to leave the line free to adjust itself, so that each turn can take its share of the strain, and only make fast the two ends of it. Otherwise one turn may take most of the strain and break, transferring the load to the next shorter one and so on, until the whole thing fails.

At sea, flags wear out quickly, so we generally take them down, as we clear the coast. But a *wind sock* does not flutter and will last a long time, so we run one of those up to the top of the mainmast, where the masthead light will illuminate it at night. A small red one shows up well against the sky.

If you have to tow another vessel at sea, use the longest line you can—probably the anchor rode—as a hawser, because she will tow better and the strain will be more even, than with a shorter one. In a tugboat we use seven or eight hundred feet, so you are unlikely to find that you have too much.

Aboard a yacht, there may be no towing bitts, so you have to contrive something to take the strain. The mizzen mast of a sailboat may be suitable—with two lines from it, to the sheet winches—or in a powerboat, you may have to carry the strain forward, on each side of her upper-works, to the anchor winch at her bow. But in any case, the hawser must be clear to pivot at a point well forward of the towing boat's stern, or she may not answer her helm properly, for geometrical reasons.

Most yachts tow badly in a seaway, due to their shape, so you need someone to steer the one you are towing. If it will take several hours, there should be a relief for him.

Chafing gear is essential—bits of old sailcloth, or even rags, will do—and when you take up the strain, be very careful not to let it happen suddenly, or something may break.

As you come into calmer water, approaching a port, you can shorten the hawser to prevent your tow from wandering all over the channel. And later, in smooth water, you may find it best to cast off the tow and have the men aboard her take in the hawser as you circle around and come up behind her, to make fast to her quarter and push her up to the dock you have chosen.

For most yachts, ordinary fenders are fine but for a very light displacement boat, we prefer to make our own, by covering inner tubes with sailcloth, then inflating them. They are light and easy to handle, they work well, and they can be deflated for stowage at sea, aboard boats where space is limited.

The ready-made fender boards you find aboard a boat are often too short to be satisfactory for mooring to piles, so we make our own (about six feet long) and varnish them to keep out water. And a heaving line with a monkey's fist on the end of it is no good, because it rolls all over the dock. So for a large boat we prefer the kind with a bag of lead shot, which stays put where it lands. But in a yacht of ordinary size, a docking line is usually light enough to heave as far as you need.

The trick to doing that is to prepare the line carefully, before you heave it. Take a fathom of it and make a smooth loop—with no kinks or twists—in your right hand. Continue until you have nine loops, then transfer three to your left hand (you lose one doing that, which leaves five in the right hand). Now check each loop, to be sure that it will slide off your fingers—endways—without getting caught in another one. Then take the right hand well back and swing it forward, letting the loops slide off your fingers when your hand stops. The ones in your left hand will follow them and, with a little practice, it is easy enough to lay the whole nine fathoms in a straight line, within a foot of where you want it.

16

Intracoastal Waterways

According to the charts, there is an intracoastal route all the way from Boston, Massachusetts to Brownsville, Texas. But it varies considerably from place to place.

The first part, from Boston to New York, takes you across Cape Cod Bay, through the Cape Cod Canal, down Buzzard's Bay, across Rhode Island Sound and Block Island Sound, then down Long Island Sound to the East River. And apart from the short canal, that is much like a coastwise passage.

The next section, from New York to Cape May, is in fact a coastwise run for most boats these days, since the waterway from Manasquan south has been allowed to deteriorate.

But from Cape May, going up the Delaware, through the canal, and down the Chesapeake to Norfolk, Virginia the water is generally so shallow that no heavy seas can devlop. And from Norfolk there is a sheltered waterway to Florida.

It was made by improving natural features, such as rivers and lagoons, then joining them together with land cuts. It is twelve feet deep as far as Fort Pierce and eight feet deep after that. And nowhere is there a bridge less than thirty-six feet wide, or a fixed one less than fifty-five feet high, until you get to Miami.

But there are many bridges, of different kinds: vertical lift types, single and double bascules, swing bridges (usually single but sometimes double), and pontoon ones. And depending on the height of your boat, some or all of them will have to open for you. So the first thing is to find out what that height is.

In a motorboat, it is seldom hard to measure from the highest point down to the waterline, but if there is anything that you can easily lower, like a radio antenna or a decorative mast, plan on taking it down as you approach a bridge that you can otherwise pass under. In a sailboat, you measure the flag halliard on the mainmast, then add whatever is above and below that, to find her height. And with any boat, you must allow for the change in her trim when she is under way.

The clearance under a bridge is shown on the chart but that is for Mean High Water, so there will usually be more, at the highest point. But there may be less, so check the gauge at water level beside the shuttering.

If you want the bridge opened, give three blasts on your horn (four in a real emergency) and see what happens.

Railroad bridges are usually left open when there is no train coming, so if you find one shut, the operator may not be able to open it for a while. But when he does, he is generally in no hurry for you to go through. So you can stay well back until you see the draw span begin to open.

At a highway bridge, the operator may not hear your horn, because of the noise of traffic going over the metal grating in the draw span. But when he does, he is likely to open it promptly and be in a hurry to close it again. So you watch the barriers go down to stop the cars and time your approach so as to arrive at the bridge a few seconds after it is open.

For that you need to know the direction and strength of the current, which you can estimate from the behaviour of the water around a pile, before you get there. And see what is on the chart, beyond the bridge, so that you will not have to look at it while you are going through. Then you can line your boat up with the center of the draw span by looking at the shuttering that protects the piers on each side of it.

As you pass through the bridge, watch out for eddies in the water and be prepared to use as much rudder as you need, to prevent her head from swinging either way. And as soon as you are clear of the bridge, give the operator a brief toot on your horn, to let him know he can close it.

From Norfolk to Albermarle Sound there are two waterways. The Dismal Swamp Canal is more convenient for sailboats drawing less than six feet, but the Virginia Cut is better for deeper and faster boats. Both have locks but they are easy to negotiate, with men to take your lines and little turbulence. And from Norfolk south, the intracoastal waterways are clearly marked, sometimes by buoys but more often by beacons and day markers, standing up out of the water on tall piles.

Those are easy to find when the background is low or distant but when there are trees close behind one, it may be hard to see and in that case you can find it by looking for the part of the pile just above the water, below the trees.

In places where there are strong cross currents, ranges are also provided and when those are in front of you, they are easy to use: You just line up the two marks and bring the boat's bow up into the current, until she is tracking straight toward them. But when a range is behind you, there is a tendency to overcorrect until you get the hang of it.

A sailboat is greatly affected by currents, because of her grip on the water and her low speed. So you have to be extra careful when approaching a bridge and there should always be an anchor ready to let go, in case the engine fails. But if her keel should touch the bottom, it does no harm and you can usually get back into deeper water before she loses headway. In fact, she will often warn you that you are straying from the deep channel, by pulling quite strongly toward it.

But in a fast motorboat you hardly feel a cross current and with twin engines you can go right up to a bridge, back her down, and wait for it to open. And they draw so little water that you should never touch the bottom, even at the shoal spots which sometimes develop in a waterway. But if you do, you will most likely bend a propeller and possibly a shaft. When that happens, you will know by the vibration and you must shut that engine down, before it does any harm.

Shoal spots develop where sand or mud is washed into the channel, sometimes by a storm but more often by the current. And the engineers are constantly dredging them out. But now and then you find one they have not dealt with yet.

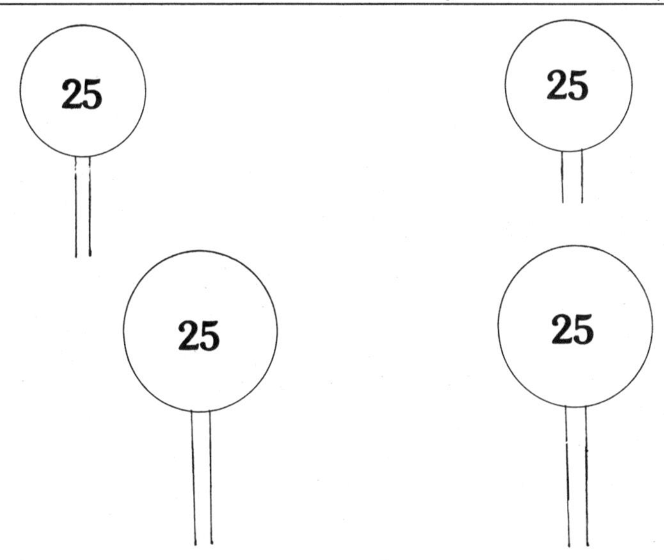

Reading intracoastal ranges: When the higher (far) part of a range is to the left of the lower (near) part, you are to the left of the range. When both parts are lined up, you are exactly on it.

You can see on the chart where they are likely to form: on the inside curve of a river, for example; or at the tip of a shoal, where two rivers meet; but especially near an inlet from the sea, where the currents are strong; and where the improved channel cuts across the natural flow of the water. So you should not be surprised when you come across one. In fact, you should be expecting it. For the secret of staying afloat in a waterway is to read the chart, all the way.

It is not enough to identify a marker, pass close by it, and aim for the next one. You have to consider where the deep water lies in relation to each marker, and how it runs between them. Is it straight or does it curve? If so, how much and where? What track must you follow to stay in it?

Then there is the current: which way is it flowing and at what speed? It changes all the time in a waterway. But you must know what heading to steer to follow the right track.

The state of the tide is another factor, for there may be more (or less) water than is shown on the chart. You can judge that quite well by looking at the banks. And back at the chart, you can figure the penalty for making a mistake.

There are very few rocks in the intracoastal waterways. Mostly the bottom is mud, up the rivers, and sand near the sea. But the sand may be hard and there are banks of oyster shells (shown on the chart) that are very hard.

Often there will be a shoal on one side of the channel and deep water on the other, so you favor the good side and if you are slightly off your track, there is no penalty.

In other places, there are shoals on both sides. And when another boat is occupying half the channel, you have to control yours with great accuracy to stay out of trouble. But there are also places where the water is deep and clear of obstructions on both sides of the channel, where you can relax and let someone else steer, until the next tight place.

Reading the chart, we stay three marks ahead of the boat. That is, we try to remember every detail of the reach she is in, while we are studying the next two. And so it goes on, all day.

The beacons and day markers are usually set well back from the channel to prevent barges from knocking them over. But now and then you find two of them, on opposite sides of it. And passing between them, you can see how

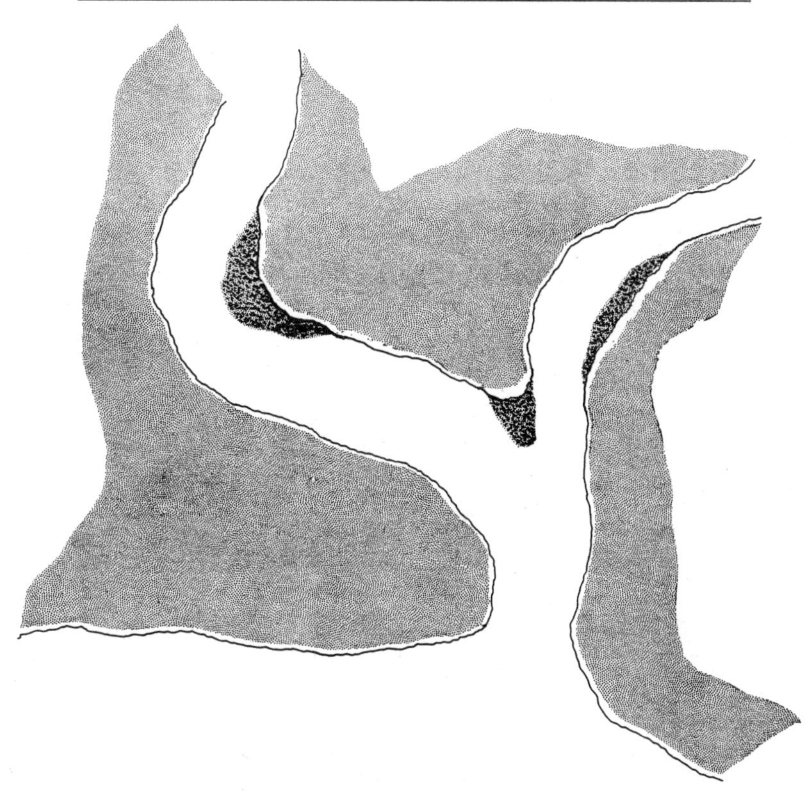

Shoals form on the inside curve of a river—down stream—and at the junction of two rivers, especially near an inlet from the sea, where the tidal currents are strongest.

far off you should be to be in the center of the channel.

At sharp corners it pays to read the compass, then add or subtract the number of degrees you have to turn (from the chart) to be sure you get the right mark next.

And your wake can be very helpful, for it will change its shape, becoming steep and even breaking where it passes over a shoal patch, and clearly marking it for you.

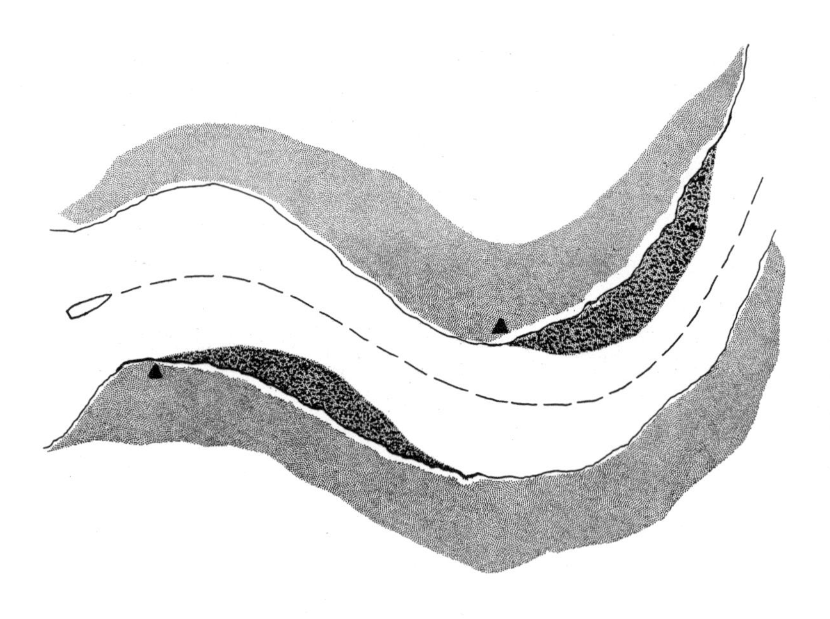

In a winding river, note the location of each shoal—on the chart—then follow a course to stay clear of it.

17

More about Waterways

As you go south from Norfolk, the waterway changes. First you have the wide, open sounds inside Cape Hatteras, connected by land cuts, as far as Morehead City. Then you are in straight, narrow channels, relieved only by the Fear and Waccamaw rivers, until you get to Charleston, South Carolina. After this, you start going up and down a series of big, winding rivers.

In that section—from Charleston to Fernandina, Florida—the channel often goes in a great curve between one mark and the next, but most of the time it is wide and deep. And then you are back in straight, narrow channels all the way to Miami.

Towboats run the waterway at night but they have powerful searchlights and their crews know every detail of the route. For other vessels night travel is not practical, except maybe for an hour at the beginning or end of the day. Then one person has to read the chart, while another follows his directions. But the banks are hard to see and the cross currents are hard to detect, so that the chances of going aground are quite high.

To make the best time, we start at first light and stop just before dark, either at a dock or in a safe anchorage. That takes some planning and we find it best to work two days ahead. Each night, we consider the next two days' runs and adjust our departure time or, if necessary, change our stopping place to be sure that we arrive in daylight.

In a fast motorboat, you have to allow for slowing down where your wake might cause trouble. But in a sailboat, the day's run will be her speed times the number of hours available, less time lost at bridges and bucking currents.

Where there is a choice, a dock generally has facilities like electricity but an anchorage is quieter for sleeping. And you can usually get fuel more quickly by stopping at a dock in the middle of the day, when no other boats are there.

But the currents are often strong near an inlet, so we prefer an anchorage that is equidistant from two of them, well off the channel, but within sight of a beacon, so that we can check any movement of the boat. And after setting the anchor firmly in the bottom, we put up an anchor light.

Where a lagoon or sound is connected to the sea by a few small inlets, it tends to stay at the half-tide level. Then the currents at the inlets flow out when the tide is low and in when it is high, regardless of which way the tide itself is going. Those are known as hydraulic currents and once you are aware of them, they are not hard to predict from the tide tables.

When you meet a towboat in a narrow place, she seems to take up the whole channel. But in fact she will generally give you plenty of room to get by. Just move over as far as you can to starboard, without going aground and keep enough way on your boat to give you positive control.

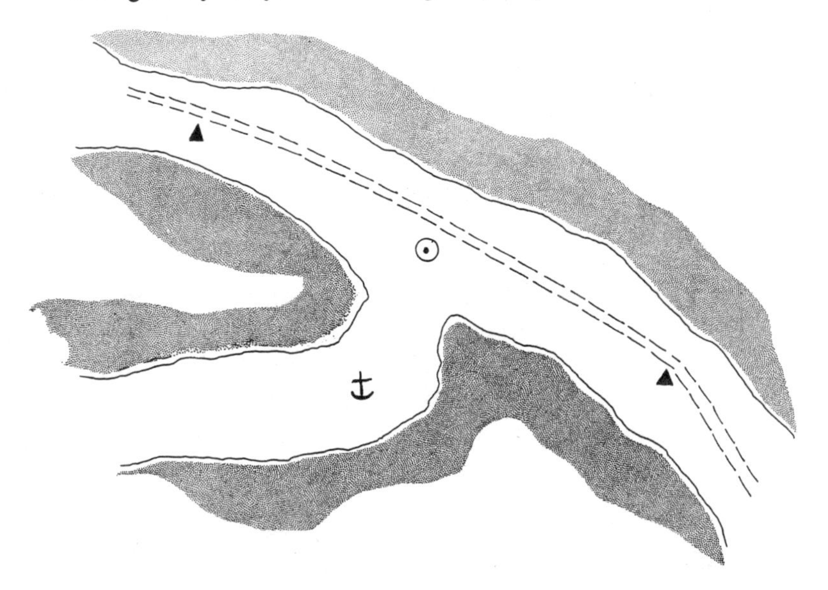

The best anchorage on a waterway is in a wide backwater—equidistant from two inlets—well off the improved channel but within sight of a beacon, in case the anchor drags.

But overtaking a towboat can be hairy, because the water she displaces rushes down her sides to fill the channel behind her. So you have a strong current against you and in a slow boat, you may not make it. In which case, you drop back and wait for a wide place, where you can pass her safely.

Fish boats are always in a hurry and in a waterway they

are often steered by young boys, while the men attend to other things. So give them as wide a berth as you can.

When you see a dredger ahead, look for the pipe along which the sand or mud is pumped away. In most cases, that forms a barrier across one side of the channel, so you will have to pass it on the other side. But if in doubt, give three blasts on your horn and the operator will answer with his whistle. One blast means you are to pass to starboard of it and two means to port. Anyway, slow right down and watch for cables leading to anchors, which are marked by barrels in the water.

Sometimes a channel leading out to sea is also a part of the waterway and then you may find a black buoy with a yellow triangle on it, meaning that it counts as a red one for intracoastal purposes. Or a red buoy with a yellow rectangle on it, that counts as a black one (the yellow color being the clue that it refers to the waterway). But if you are reading the chart as you should, it will not confuse you.

In Florida there are bridges which are closed at certain times, so you have to find out when the ones ahead will open and plan your day's run to make the best of it. Inside Cape Kennedy there are some that are closed during the morning and evening rush hours. And farther south, from Palm Beach to Miami, there are many that only open at specified times, each hour.

So if the weather is good, it may pay you to go outside at Fort Pierce or Palm Beach. In which case, you stay inshore to avoid the Gulf Stream, going south. And there is a fixed bridge fifty feet high over the waterway, just before Miami.

Palm Beach is a good port of departure for the Bahamas, since it is an easy passage from there to West End, where you can get clearance to proceed through the islands.

South of Fort Pierce, there is another waterway that goes across Florida via Lake Okeechobee to Fort Myers. But there is a fixed bridge forty-five feet high, just before the lake and there may be as little as four feet of water in some places. It has locks in it but they have no sluice gates, so the operators open the main ones to let the water out. And going upstream, it pays to stay well back from the gates while they do that.

There is also a waterway south of Miami that takes you around to the west coast of Florida by a comparatively sheltered route. But there again, it is only reliable for boats drawing four feet or less and otherwise you have to go outside.

From Fort Myers you have three short coastwise passages, to Clearwater, Cedar Keys, and Apalachicola. But then there is an intracoastal waterway much like the east coast one, all the way around the Gulf of Mexico to Brownsville, Texas.

Boats with masts over fifty-five feet high can use the waterway between Norfolk and Morehead City but they have to take the side channel at Beaufort to avoid the fixed bridge.

And you can still run the waterway from Manasquan to Cape May in a motorboat (less than twenty-five feet high and four feet deep) but it is much narrower than the others and south of Atlantic City you need a couple of feet of tide under you.

The Cape May Canal, from the harbor into the Delaware, is useful if you can clear the fifty-foot fixed bridge, because the shoals off the cape are always shifting and the only other safe route is via the ship channel, across the bay.

The Chesapeake and Delaware Canal and the Cape Cod Canal are both ship canals, wide and deep, with riprap along their banks and strong currents running through them. But the traffic lights are for the ships and you may go through at any time, so long as you stay out of the way. Just remember that a ship's stern swings out as she turns and if you meet one at a corner, try to pass her on the inside of the curve.

In winter, there may be ice on the waterways as far south as Morehead City, which would prevent a wooden boat from moving (it would cut her in two) but is seldom thick enough to bother a metal or plastic one. And though it is cold, running all day, an electric heater plugged into a dock will keep the cabin warm and dry at night. So there are few days in a year when you can not make good progress toward your destination.

18

Electrical Systems

Ashore, most things that run on electricity are very reliable. Lights, refrigerators, fans—things like that—go on and on, year after year, working when they are wanted and lying idle when they are not, with hardly any attention.

But in a boat, it is different. Electrical things fail more often than any others. Why is that?

The high humidity in the air near the water has a lot to do with it, for moisture finds its way into the works and, being impure, it provides paths along which electricity can go where it should not. And any salt that may be deposited will absorb water from the air, so that it never dries out.

A small leak in the boat's upperworks can put a delicate piece of electronic equipment out of action and in a storm at sea, half a ton of water coming down the main hatch will usually leave you without electricity for the rest of the passage.

In fact, having to make and store the electricity, as well as use it, is half the problem. But there are two cases where that does not apply: when things are run off a shore line and when they have their own dry batteries.

With a shore line, you can run heaters, air-conditioners, cooking stoves, and other things that need a lot of electricity. And since they are quite simple, they usually work. So when you are at a dock, you can have all the comforts of home, without too many problems. And why not?

But most reliable of all are the things that run on dry batteries, especially if they are made for use at sea. We have had waterproof flashlights that worked for years, in all kinds of weather, with very little care. And ordinary radio direction finders, which also get weather reports and time signals, rarely fail (though we try to keep those dry). In bad weather at sea, those are the things that you can trust to work, whenever you need them.

But between the two extremes—a boat tied up to a dock and one in a howling gale, far from land—lie the conditions in which most of us operate, most of the time. And for those, each of us has to find his own compromise.

The more complex we make our electrical system (or allow it to become), the less reliable it will be.

There is a theory that says reliability can be gained by having two (or three) of everything, with all sorts of circuits to feed one thing from another. And in commercial vessels, or in large yachts with full-time crews, it works. But the maintenance is endless and for most people that means finding electricians, waiting while they fix things, and paying large bills.

So the simplest system that meets your needs is generally best. But what items you can have, without undue problems, will depend to a great extent on how you use your boat.

For offshore cruising in a sailboat, there should be one battery that does nothing but start the engine and can not be connected to anything else, except its own alternator. Sometimes you find the navigation lights hooked up to that battery; if so, they should be taken off.

Then there should be a second alternator, driven by the engine, that charges a twelve-volt battery (or two of them, if you need more capacity), which feeds everything else.

Each circuit should be clearly marked on the fuse panel, with its own switch. And there should be an ammeter to show the current drain. Then if a leak develops, you can throw each one in turn and find out which circuit it is in.

Most people prefer not to run the engine for more than an hour a day at sea, to charge the batteries and keep it in shape, so you can only have things that take little current and ideally you should stick to those that almost always work, which means a bilge blower, a bilge pump, navigation lights, cabin lights, and an echo sounder (for making landfalls in thick weather).

The cabin lights should have small bulbs. (Bright ones not only use too much current; they make you sick.) And the masthead light should have two bulbs, with two switches, so that when one of them goes out at sea, you can turn on the other one without having to climb the mast and change the bulb.

You could also have Loran (it does not use much current) but it is not as reliable as the other, simpler things.

For your other needs, you have to use things that do not take electricity, like an ice box, an alcohol stove for cooking, a charcoal stove for heating, and so on.

And even with such a simple electrical system, you need backups, like a starting handle for the engine, a manual bilge pump, and another one that runs off the engine.

Also, it is possible for a short circuit to start a fire with a twelve-volt system (it has happened to us), so you must be able to isolate the battery in a hurry.

For coastwise cruising, where there is no need to be out in storms and most people use the engine for a couple of hours a day, you can reasonably have more electrical equipment. Spreader lights are handy and Omni is great, when there is a station within range. Brighter bulbs in the cabin are nice to read by in port. And a capstan to get the anchor up saves a lot of work (though we run the engine when we use that).

Radiotelephones can be useful for making phone calls but it would be most unwise to rely on one for communication in an emergency, since the conditions that produce the emergency are quite likely to put the radiotelephone out of action.

A pressure water system, which we would not use offshore (for fear of losing our water supply), is all right for coastwise cruising and if you add a water heater of the kind that works off the engine, you can have hot showers.

In a motorboat, where the engine is going all the time you are under way, you can have radar and an autopilot, if you really need them. But there you are getting into more complex equipment, which can more easily go wrong.

And with an auxiliary generator, you can have anything you want, including a refrigerator, a freezer, electric heating, and air conditioning. But the more things you add, the higher are your chances of spending your time looking for an electrician to fix something, instead of enjoying your boat.

In any case, be sure to keep the owner's manual for each

piece of equipment in a safe place, with an up-to-date circuit diagram of the whole system. And if possible, you should be able to handle small problems yourself. For that you need spare bulbs and fuses for everything, spare wire, and electrical tools. But most important is a testing meter, without which it is hard to find out what is wrong.

19

Of Anchors and Dinghies

It is often more pleasant to spend a night at anchor than at a dock. But to be safe and comfortable, you have to pick a good spot and make sure that your anchor will hold.

Most of the information you need to select an anchorage is on the chart, so we study that before we get there, looking for places that are sheltered from the weather, clear of any traffic (as under the lee of some trees, in the bight between two beacons), where the water is deep enough but not too deep, and the holding ground is good. Ideally, the place should also be out of any strong current and easily identifiable.

There is an old adage: Never anchor in an open harbor in winter. This makes a lot of sense, for in winter the weather systems move much faster than they do in summer, so your chances of being caught by a sudden change are much greater, while the change itself is likely to be more severe. But even in summer, we avoid open roadsteads (unless the weather is very stable) and look for a place where there is land on every side, so close that no matter what the wind may do, it can not create enough of a sea to bother our boat.

The bottom should be sand or mud, rather than gravel or rock, but watch out for places where the chart says it is "sticky" since in those spots your anchor may pick up a great lump of mud and slide along the bottom, doing no good at all.

As long as there is good shelter and the bottom is soft, it need be no more than a foot or two beneath your keel at dead low water, so that mud flats often make excellent anchorages, especially when there are buoys or beacons making an adjacent channel, enabling you to check your position at night. And if there is a current, it will be weakest there.

When approaching a strange anchorage, it pays to have some daylight to spare, in case the one that looked best on the chart has a snag (such as a fish trap in the middle of it). And sometimes the chart is quite wrong about the depth of water in an out-of-the-way place like that. So you should take soundings wherever your boat can swing before anchoring there.

Having picked the best available spot, we get the anchor ready and check the wire on the shackle pin. Then we haul some rode (about three times the depth of the water) up on deck, coil it down, turn the coil over so that it runs off the top, and make sure that it will lead clear through the bow chock without fouling anything when we put the anchor over. In a larger vessel, which uses chain, we send a man below to make sure that everything is clear in the chain locker, then lower the anchor to the waterline.

Next we head the boat into the current and bring her

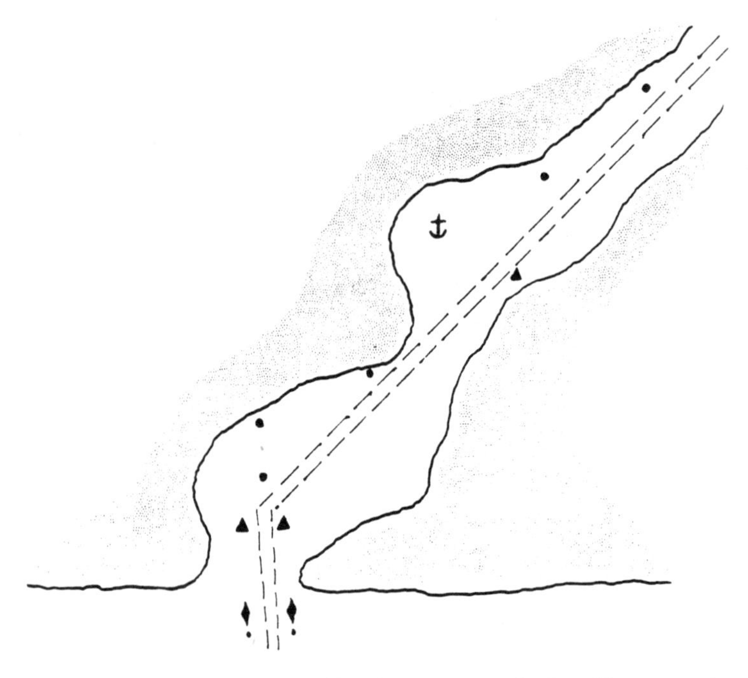

An anchorage in a harbor like this is easy to find, well-protected, and clear of river traffic. Note that the front part of the range at the harbor entrance is also a port hand channel beacon.

to a dead stop (over the bottom) in the spot we have chosen. To do that, we watch the details of the shore alongside us—those near the water's edge and those farther back—until we can see that we are no longer moving forward (though we may still be going ahead through the water). Then we drop the anchor carefully over the side, so that it goes straight down, trailing the rode behind it.

By that time, the boat should be drifting backward on the current and when the anchor hits the bottom, we grasp the rope and pull it, to bring the anchor into line with it. Then we pay out the rode as the boat falls back, keeping just enough tension on it to stop her bow from falling off to one side.

The amount of scope we use depends on the circumstances but generally it is about a fathom for every foot of water and when that is out, we snub the rode and see how it behaves. It is best to snub it gently, so that the anchor digs in rather than snatches out of the bottom. And with a hand on the rode, you can feel what is happening. Often it will catch, then let go, drag for a moment, and catch again. But once it seems to be holding, you can signal to the man at the wheel to back down on the engine and set it firmly in the ground.

That means going astern and watching the shore until the rode is taut and the boat is no longer moving backward, then gradually increasing the revolutions—still watching for signs of further movement—until he is sure that the anchor is well and truly embedded in the bottom.

When he stops the engine, the boat will come ahead but by the time you have put the anchor light up on the forestay, she should have settled down and you can take the bearings of some marks on the shore, in case she drags the anchor later.

If there is no current, we follow the same procedure but we head into the wind and go astern to pay out the rode. And if there is a strong wind against the current, we decide how the boat wants to lie and set the anchor that way.

For sand or mud, we like a Danforth anchor; properly set, it hardly ever drags. But in gravel, a stone can get wedged between the flukes and the shank, so that when the tide turns, the anchor pulls out. And with rock, it is a matter of hooking onto the bottom, rather than digging in. So for gravel or rock we prefer an old fashioned anchor. But in those cases, we set an alarm clock and go on deck at the turn of the tide, to make sure that the anchor holds securely in the new direction. When you can not see any marks, you can still tell if the anchor is dragging by dropping a lead line onto the bottom near the boat and seeing if she moves astern, past it.

But usually the wind dies down at night and then the boat lies quietly to her anchor on the still water, with just a faint chuckle of water past her hull. And in the distance, you can see the lights of houses on the shore.

In the morning, we take in the anchor rode until it is up and down, then snub it and go ahead with the engine to break the anchor out. And for coastwise cruising, we often stow it on deck. But for an offshore passage, we stow it below—under the cabin sole, if possible—to leave the foredeck clear.

A short length of chain between the rode and the anchor increases its holding power and the plastic-covered ones that are sold for the purpose are nice. But for anchoring on coral, you need a longer piece of chain, since the coral would cut the rode if you allowed them to touch.

If ever you run aground, the instinctive thing to do is back down on the engine. And with a motorboat that often works, because her engines are powerful and the surge of water from her propellers washes the sand or mud away from the boat. But with a sailboat it seldom works, since her engine has less power and her propeller is inefficient for going astern. So the way to get her off is to use the anchor.

First you take the dinghy and find out where the water is deeper. Then you put the anchor aboard the dinghy and coil the rode down between your feet, leaving just the end made fast to the yacht. Then as you row off toward the deeper water, the rode will run out over the stern of the dinghy, with very little drag. And when it is all gone, you drop the anchor.

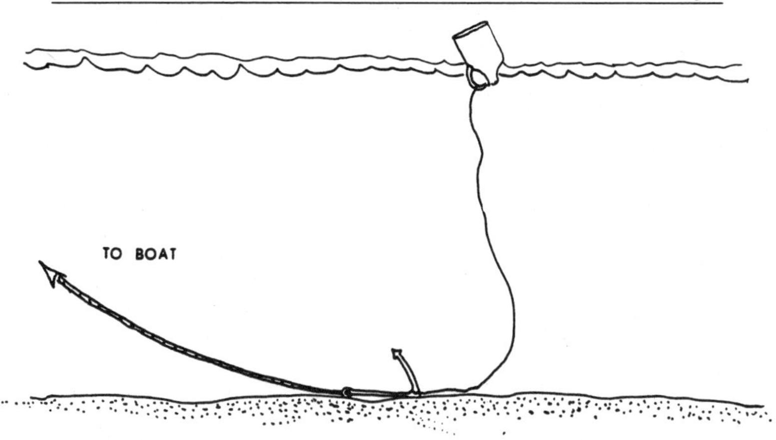

A plastic jug makes a fine anchor buoy. A line from it to the crown of the anchor allows you to trip the anchor, if it is wedged between two rocks when you want to leave.

Back aboard the yacht, you take the end of the rode to her bow—using the slack as the anchor digs in—and lead it to a winch. If there is no winch forward, you rig a snatch block and take the strain aft to one in the cockpit, but the pull must come from the bow, because the object is not to drag the boat off the bottom but to turn her, which takes far less power.

As you take the strain on the rode, the boat starts to pivot and you put the engine full ahead, with the rudder hard over, which increases the turning moment and twists her loose. Then as she begins to move forward, you straighten the rudder—keeping the rode taut, all the time—and with any kind of luck she will slide off the shoal into the deeper water.

When rowing a dinghy into a strong wind, it pays you to feather your oars. (You can feel the difference in the drag.) If, when going ashore, there is a strong current against you, it pays to go out of your way to avoid it. But when you get there, the question is often what to do with the dinghy.

At a commercial dock, it is usually best to tie it fore and aft across a corner, where it can not get under the pilings or otherwise destroy itself. And you can carry a light dinghy up a beach, beyond the high water mark. But with

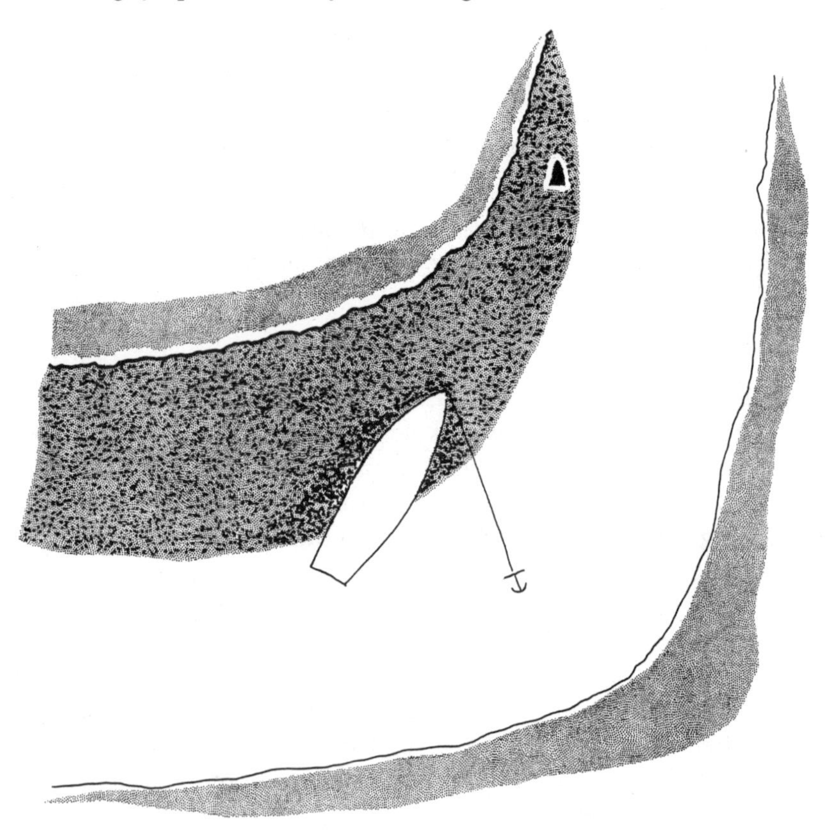

To get a sailboat off a shoal, put an anchor in deep water abeam of her, take the strain on the rode—to twist her loose—then go ahead with the engine, to slide her off it.

a heavy dinghy, it is best to use a running mooring in a beach situation.

For that you attach a block to a light anchor and through it you reeve about twenty fathoms of line, with both ends made fast to the bow of the dinghy. Then as you approach the beach, you drop the anchor and after you disembark, you can haul the dinghy out, so that it lies to the anchor. A peg in the sand will hold the line and when you come back, you can haul it in again.

Most dinghies tow much better from a point just above the waterline than from higher up, and on a long painter rather than a short one. And they generate less drag when riding on the front of a wave, than they do on the back of it.

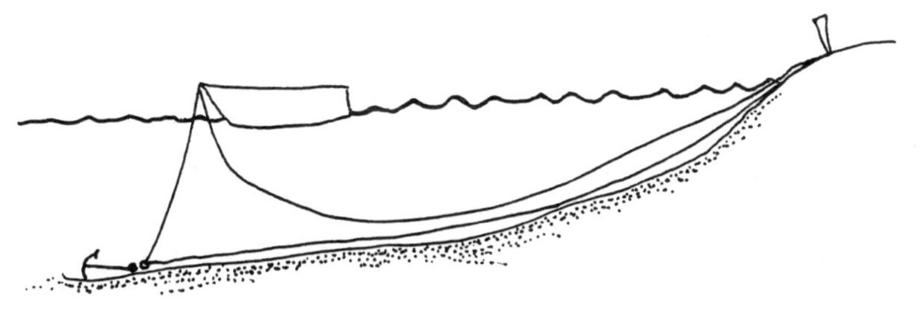

The dinghy is lying to an anchor, some distance from the shore, by a continuous line that goes from her bow, down to a block attached to it, and back to a peg on the shore. The other part of the line is used to haul the dinghy to shore.

But for coastwise cruising, we prefer to have the dinghy in its chocks on the coachroof and for an offshore passage, we often secure it upside down on the foredeck, where it is less likely to be taken off by a wave. In which case, we rig a life line from the yacht's bow to her mast, over the dinghy, to hang onto when we are working on the foredeck.

*Current and Leeway**

In a ship or a fast power boat, it is enough to keep a dead reckoning (DR) plot on a basis of course and speed, without regard for current or leeway. But since the increase in the cost of fuel, few fast boats go far. And for a slow one, or a sailboat, such a plot is quite inadequate, for the current may at times be going as fast as the boat, in a different direction altogether.

So we assume the reader is using a more sophisticated kind of dead reckoning, including current and leeway, as many people have been doing since the 1930s. But in case anyone has not yet tried it, the following notes may be helpful.

Leeway can be as much as ten degrees in some cases and six is common, for a sailboat going to windward. It is easy to find, in slack water, by passing close to a mark and taking back bearings as you go away from it. And after a while, you can estimate what it will be, in various conditions, to within one degree. Then you simply apply it as a correction to your heading.

The direction and speed of the current are easiest seen on current charts—when they are available. Where two figures are given, the higher speed is for spring tides and the lower is for neaps, so you must interpolate for the date.

When there are no current charts, tables can be used, with ordinary charts, to estimate the current at any time

*This section on Current and Leeway is an introductory note to prepare the reader for the discussion in chapter 20 on dead reckoning.

and place, for it is not hard to imagine how the water must flow—around a headland, across a bay—if you see the geography.

When you pass a buoy, you can tell how fast the current is going and if you take a back bearing when you are well away from it (and know your leeway), you can tell its direction. So after a while, you should always know what it is doing.

Offshore, we use the current shown in a pilot chart (which is an average figure) but adjust it when a run between two fixes tells us it is going faster or slower than usual.

To find your position, you lay off the course and distance from the last one—as if there were no current—then add it by drawing another line, from the end of the first one, to show how far and which way the water moved in the same time.

Going to windward in a sailboat, we do that each time anything changes. But off the wind or in a power boat—when we can choose which way to go—we allow for the current, when figuring our heading. And then our DR position, at any intermediate time, is a proportion of the distance to the next mark.

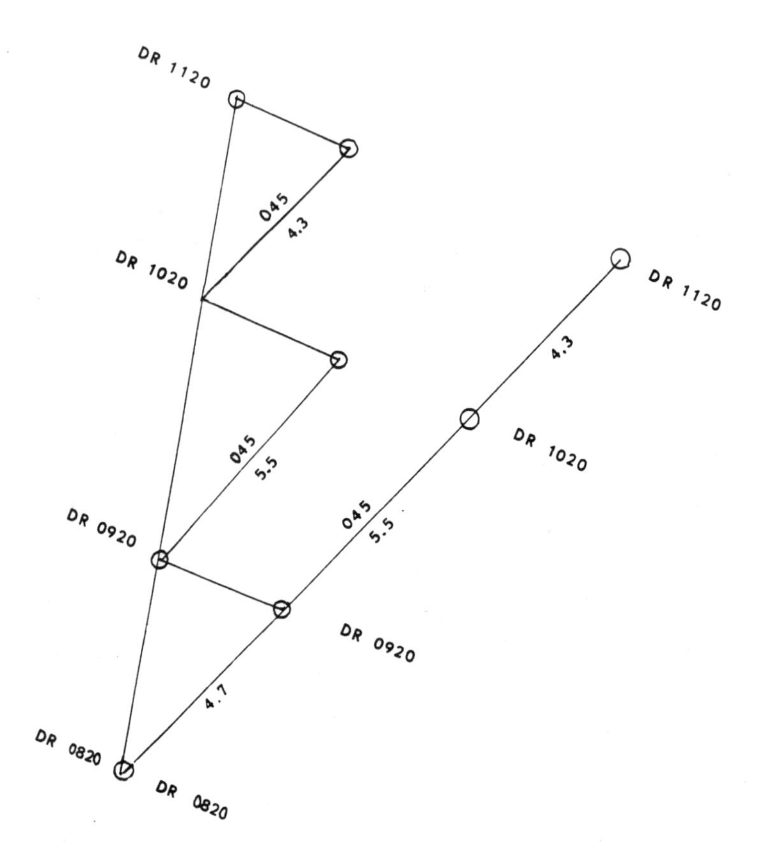

Allowing for a current: A boat is going northeast at 4.3 to 5.5 knots. The current is going WNW at 2.5 to 3.1 knots. The plots on the right show DR positions, without current. Those on the left include allowances for it and the last one shows that the boat is in fact about 8.5 miles from where the other method indicates.

20

Dead Reckoning

Dead reckoning (DR) is the most useful kind of navigation. It is very flexible—suitable for a run of a mile across a foggy river or hundreds of miles across open water—yet the things you need to do it are so simple, they rarely fail. And even if one does, all is not lost. Just the accuracy decreases.

It can be used to fill in the gaps between positions you get by other means—celestial, electronic, or visual—and to provide a transition from one kind to another. But when these other means fail—the sky clouds over, the electrical system quits, or fog rolls in—dead reckoning still keeps right on working.

Basically, it only requires that you keep a careful note of how far you have come and in what direction, since last you knew where you were. Then by simple geometry you can find out where you are. But the best way to do it depends on how far it is between one fix and the next probable one.

Before crossing a river in fog, you must know the heading, distance, and running time to the mark on the far side. But there is little need to keep a record of such a short trip, so you must remember the time, as you leave—in minutes, past the hour, like "06" or "33" or "51"—and you can easily figure in your head when to look out for the mark.

But if there is a deep channel in the river and you meet a ship, you may have to change course or stop to avoid her. Then your original plan is no longer valid but by using mental DR you can figure where you are and how to get to the mark.

Just by nothing the time, you have a rough fix. You are 40 percent—or whatever—of the way across. And the best thing is to stay there, while the ship goes past, by heading into the current and going at its speed. Then you can note the time and continue with your original plan, as soon as the ship has left.

But if you have to change course, you can still keep track of where you are by noting your heading and how long you are on it, then figuring how far the water has moved in that time. And from your new position (after the flap is over) you can figure a heading and time to the mark, in the usual way.

There is no need to go groping toward the far bank, for it is quite possible to continue on your way. But dead reckoning is an art and to become good at it requires practice.

A fine time for that is on a coastwise passage, in clear weather. You figure your heading and time to a distant mark in the usual way but when you see it, you do not change course to go there. Instead, you stay on your heading, run out your time, and see how close you came. Then you figure what went wrong to cause the error and the next time you should do better.

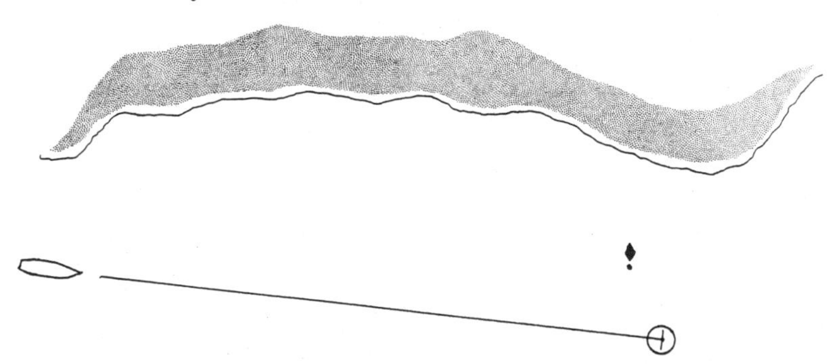

To improve your skill at dead reckoning, figure your time and heading to a distant mark but when you see it, do not go to it. Instead, run out your time and see how close you came. Then figure what went wrong to cause the error you find.

The accuracy you can hope for depends on many things —the kind of boat, her speed, the weather, and so on— but if you come within a tenth of a mile of your target, for each hour since you started, you are not doing badly. And knowing what kind of error to expect, you can tackle more difficult runs. For example, you could sail a series of courses—out from the coast, along it, and in to your destination—without taking any bearings along the way. Then you can see what your error is and figure what mistakes, if any, you have made. That way, you can improve (and verify) your skill at dead reckoning, before using it where nothing else is available. But to increase your accuracy, you have to be more precise and take more factors—or corrections—into account.

You need to estimate your speed to one tenth of a knot

and steer your boat to the nearest whole degree (beyond those levels the likelihood of inducing errors outweighs the advantage of the increased precision). And you need to allow for new things, like surface drift and the different helmsmen's errors.

We assume that you are already allowing for the ordinary currents in the water at various points along your route, and for the leeway induced by the wind under different conditions, and for the variation and deviation of the compass.

Surface drift occurs when the wind has been blowing from one direction, for several hours or even days. Then the top of the water—maybe ten feet or so—may start to move at a speed as high as a knot, downwind. That is over the main body of the water, so you have to add the two geometrically.

The helmsman's error varies with each person, so you have to watch them steer. Usually they err upwind in a sailboat, for they let her come up, more than they bear off. If there is any fear of gybing, they do it even more. With a powerboat it tends to be similar going downwind, but going to windward helmsmen often err the other way, because a sea will push her off course and they do not correct for the yards they have lost.

The error you find, the next time you get a fix, will let you know how well you are estimating the new corrections. Then it takes practice, to refine them. But on a long run, the small residual errors in all the different corrections often tend to cancel each other out and you can make a landfall after several days (using DR alone) that is surprisingly accurate.

Approaching New York out of St Thomas in *Wind Song,* we had no electronic equipment aboard and for the last two days there was no hope of taking a celestial observation in a rising gale. Yet we came within one hundred yards of our target—buoy B, about twenty miles southeast of the Ambrose lightship—in driving rain with very poor visibility, entirely by dead reckoning. Of course it took twenty years' experience to reach that level of proficiency but every small improvement helps and practice is the key to steadily increasing your accuracy at it.

The basic tools for dead reckoning are a notebook, a watch, and the boat's compass, none of which are likely to fail. But we take a spare compass and another watch, just in case. And for an offshore passage, we take a patent log, as well. That may not work all the time, because a shark can bite off the spinner (we carry spares) or it can be fouled by weed. But while a log works, it is a great convenience. It has to be calibrated, like any navigational instrument, but that is not hard to do. You make a few runs, allow for the currents in the water, and find its error, as a percentage of distance.

A boat from the Great Lakes may have spinners marked *S* for statute miles and if you use one of those at sea, you have to convert the readings to nautical miles when calibrating the log. But runs between celestial fixes will do for that, if you do not have time for others before going to sea.

After many years, dead reckoning of a kind can be done in your sleep. Your subconscious mind goes on making the necessary calculations and you wake up knowing where you are, even if you do not know how long you slept. This is quite useful. But the trick is to know when to trust it. A good scheme is to heed any warning it gives you but not become complacent when things seem okay; make all the usual calculations, anyway.

A specialized form of DR is "prenavigation," where you do the calculations before you need them. That is useful at times, when you are short handed, in heavy weather. Let us say that you plan to make a landfall far from your port of arrival and there is a dogleg (or something to go around) between them. Then it is a good idea to figure the heading and time to the turning point plus a selection of headings and times to the port, depending on what error, if any, you find at the turning point.

Then if the current or some other factor is not as you had expected, you can check the error at the turning point, refer to your figures, and head for the port, without having to go below—in small, congested waters—and refigure everything.

Sometimes, when you are approaching a landfall, you do not know exactly where you are (for whatever reason) but there are a couple of tricks which can make it relatively safe.

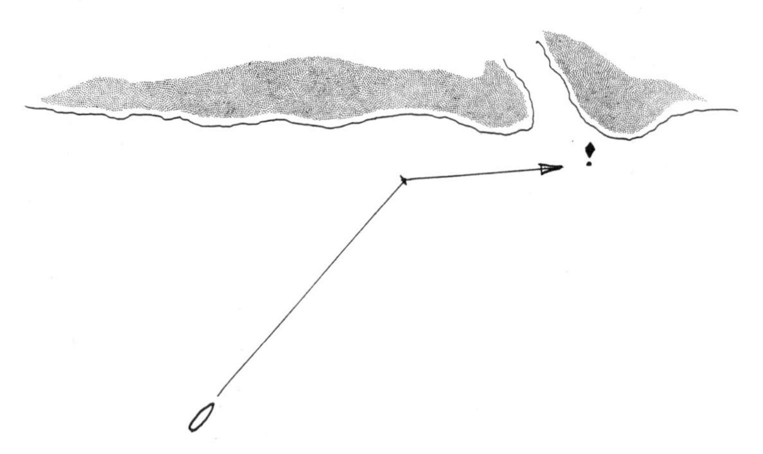

Making a landfall to one side of your port of arrival will eliminate any doubt as to which way to go along the coast.

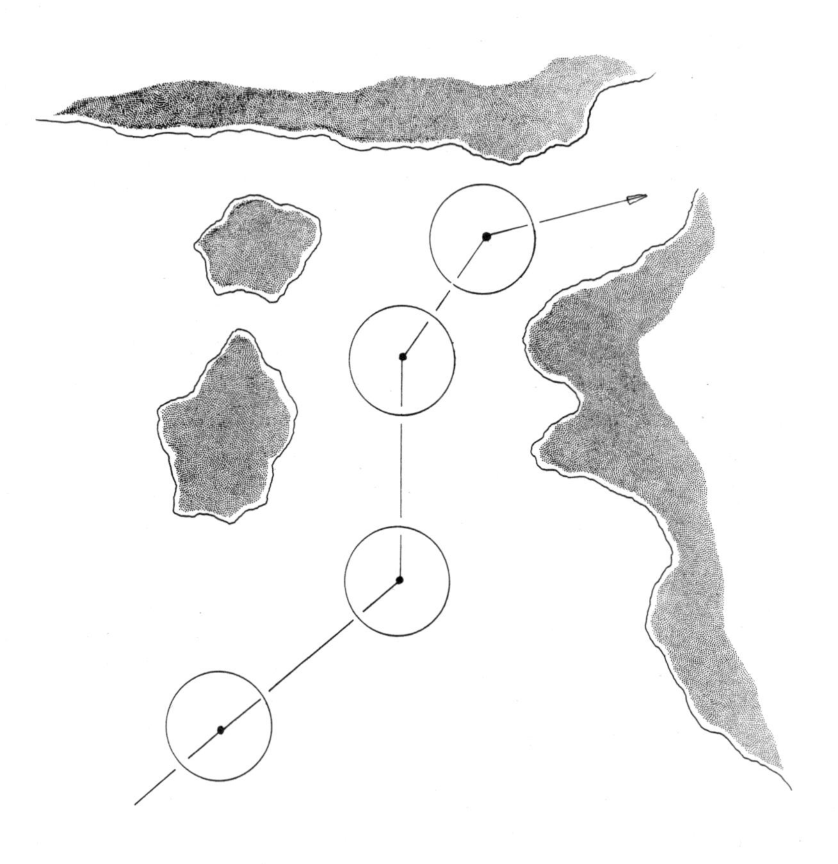

If you are unsure of your position when making a landfall, place a coin on the chart to cover the area you might reasonably be in, then navigate so as to keep it clear of danger, until you come into protected water or up to a safe landfall.

One is to figure how far you can reasonably be from your plotted position—say five miles—and place a coin on the chart which covers the whole area you might be in. Then you navigate so that the coin stays clear of all dangers until you bring it into protected waters or up to a good landfall.

Another is to aim for a point on the coast a little to one side of your port of arrival (if the coast is safe) so that you know which way to turn, when you come to the shore. Then, having studied the chart ahead of time, you are all set to pick up your marks and refine your position, as you go on in.

But more often, there is a gradual transition from DR to pilotage as you approach a landfall. First, the radiobeacons become clear and useful (about sixty miles off), then you begin to see ships, whose tracks you may know. And with luck, you will come onto soundings, where the depth finder is helpful. But do not stop your dead reckoning until you are sure that you know where you are, for it would be embarrassing to get lost at the end of a passage, outside your port of arrival.

21

Making Port

An hour or so before you reach your port of arrival after a passage, it is a good idea to show your crew a chart of it and tell them exactly what to expect: which way you will go in, what dock you will use, how you will approach it, what the conditions of wind, tide, and current will be, at that time, and what each crew member will be expected to do as you dock. If you do that well, there will be no need for shouting at the last moment. In fact, more often than not, the procedure of docking can be accomplished without a word spoken.

Meanwhile, the next step is to make sure there is no doubt about the identity of the port, or the way into it. Nothing less than certainty on those points will do, and every available clue should be checked against all the others, for you will be coming into narrow waters, where small errors are critical.

As you approach the entrance, it pays to study the chart —really hard, this time—until you know by heart where each mark you need is, what it looks like, and the channel between them. By making a real mental effort, you can retain all that information for the short time you need it. And doing so will leave you free to watch for other things, like ships, along the way. That technique is especially useful at night, when you can memorize the "code" of each navigational aid that you need—the color and frequency of its light—and your heading, from one to the next (with allowances for currents, of course).

Then as you go in, you check the current to be sure it is running as you expected and look out for anything that may have changed since the chart you are using was made. Sometimes you have to change the plan for docking the boat—because of unforeseen circumstances—at the last moment. Then it is best to stand off (usually heading into the current) and tell your crew the revised plan, before approaching the dock.

Such plans vary greatly but in a simple case, where it is only necessary to come alongside and make her fast, two people are usually enough, for a boat up to sixty feet long. We will call them the skipper and the deckhand, for reference.

As you approach the dock, the deckhand hangs four fenders and two fender boards over the side, spaced so that the centers of the boards are the same distance apart as the pilings of the dock. Then he (or she) takes a long docking line, makes one end fast aft, leads it out through a chock, brings it forward (outside everything), and coils it down, near the skipper.

Next, he takes another long docking line, makes that fast at the bow, leads it out through a chock, and brings it aft to a point from which he can easily step ashore.

Then the skipper brings the boat to a dead stop, with

the two fender boards a few inches from two pilings, and shuts down the engine, as the deckhand steps ashore, walks forward, puts a hitch over a bollard, and leads the rest of his line aft, to use as a spring line. Meanwhile the skipper steps ashore—with the other line—walks aft, puts a hitch on a bollard, and leads the rest of his line forward, to be the other spring.

The whole procedure scarcely takes ten seconds, from the time the boat stops, until she is secure, so she has no time to go anywhere. Before the people on the dock realize what you are doing, it is all over and you are back aboard.

That is an advantage, because strangers—with the kindest intentions—sometimes do weird things, like pull the boat's bow in toward the dock, thereby causing the current to slam her into it and taking all control away from the skipper. So we try to avoid accepting any help from a person on a dock but if we do, we toss him a line with a bowline in it and ask him to place it on a specified bollard or cleat. That may be necessary when docking in a slip, where four lines have to go in different directions. But even there, you can often handle it, by putting out two lines on the side from which the current (or a strong wind) is coming, then slacking them off, until you can reach out to attach the others.

Docking under sail takes more people—unless the boat is small—but usually it comes at the end of an offshore passage (because the engine was washed out) and there are three or four aboard, so it is just a matter of organizing them.

The main problems are that your power supply—the wind—is likely to vary suddenly and you have no "brakes." One way to cope with the former is to take down everything but the mainsail, then have one man raise or lower it—on your signal—to increase or decrease its size. And to stop the boat, you may have to use a long line, taken ashore in a brave leap by a deckhand, then surged around a bollard or cleat.

Coming into Bermuda with the fifty-two-foot yawl *Maribel,* out of Barbados, we had to dock in a tight place, so two people—one my wife, June—leapt ashore and with lines at both ends of the boat, they brought her very neatly to a gentle stop.

A single-engined powerboat is much like a sailboat, when it comes to docking, except that she has a rubbing strake which usually eliminates the fender problem—and the need to dock at one exact place. But a twin-engined boat handles so well, there is no problem at all. She can go ahead or astern, at will, turn in her own length and even go sideways, if needed.

To walk her sideways, you put the wheel over and go ahead with the engine on that side, while you go astern with the other one, using enough power to stop her from creeping ahead. Then the water passing the rudder (from the engine going ahead) tends to push her stern sideways, while the turning effect of the two engines pushes her bow the same way, so she goes sideways in the opposite direction from the way you turned the wheel.

How well a boat does that depends on the distance between her propellers, the size and location of her rudders, her hull form, and other factors, so it is best to practice it, in a wide place, before using it. But there is a simpler variation of the same principle, which presents no problems.

When approaching a dock in a cross wind, we often use the engine on the lee side only, correcting with the rudder to make her track straight. That induces a lateral force

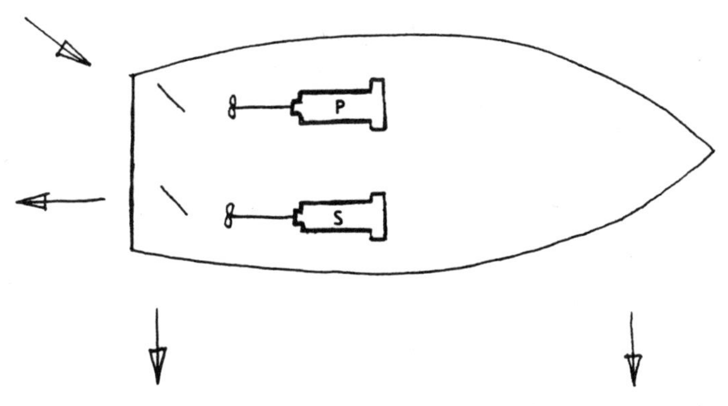

Walking a powerboat sideways: With the wheel over to port, the port engine (going ahead) tends to push the boat's stern ahead and to starboard, while the starboard engine (going astern) tends to push her astern. The net result is a force tending to push her stern to starboard. At the same time, the twisting effect of the two engines (one going ahead, one astern) tends to push her bow to starboard. So if the throttles are set properly, she goes slowly sideways.

which tends to counteract the effect of the wind on the boat, so that she will handle as though it were a great deal less strong.

Docking a small boat is seldom hard, because she can turn so quickly and you can stop her with a hand on a piling. But to dock a big one, especially a big sailboat, requires preparation if it is to be done reliably and without fuss.

At one time, we had to dock the 110-foot schooner *Azara* at Fort Pierce each day, after research operations. The dock had a large overhang, so first we got some tractor tires and hung them in two places, where our fender boards would come. Also it was across the end of a channel, so we had to turn her through ninety degrees, then lay her alongside it. But the wind in the afternoon blew strongly onto the dock and would slam her into it, unless we brought her in under control.

So each day, as we came in from the sea, we put the whaler overboard—slowing down, in the calm water—and sent it ahead of us, with three men and a long docking line. Then we continued down the channel, while other men put out the fenders and their boards, coiled down more lines where they would be needed, then got a small (600-pound) anchor ready to let go.

About a mile from the dock, we closed the throttle and 400 yards off, we took the engine out of gear, letting her slow down until she barely had steerage way. Then we put the wheel all the way over and gave her a short burst of power, to start her turning, without going ahead much faster. When she was about forty-five degrees to the dock, we would signal to a man standing by the anchor, to let it go. Then as he veered the chain, *Azara* would swing parallel with the dock.

Meanwhile the whaler had put two men on another dock, the same distance from ours as was the anchor. One payed out the long line, as the whaler came back to Azara's stern, while the other walked around to our dock, ready to take our lines. As soon as the end of the long line reached our stern, we were safe. Then the wind could not possibly slam *Azara* into the dock and we could "lower" her gently toward it, by slacking off the anchor chain and the docking line, together.

Centering the rudder, we used the engine to move her fore and aft, so that she landed exactly on the fenders. Meanwhile the whaler came around to the offshore side and lay a few yards off, ready to give her a nudge with its bow—to straighten her out—if we pointed to a place it should push. As we closed the dock, the other lines went ashore and as we came alongside, the gangplank was run out. In a few seconds the engine was shut down and it was all over, with no shouting or fuss. And it went like that, day after day.

Of course we had twenty people aboard but only five of them were our crew; the rest were scientific staff. The secret was in the planning, telling each person what to do, plus using the things we had—like the whaler—to the best advantage.

When you arrive from another country, you must clear

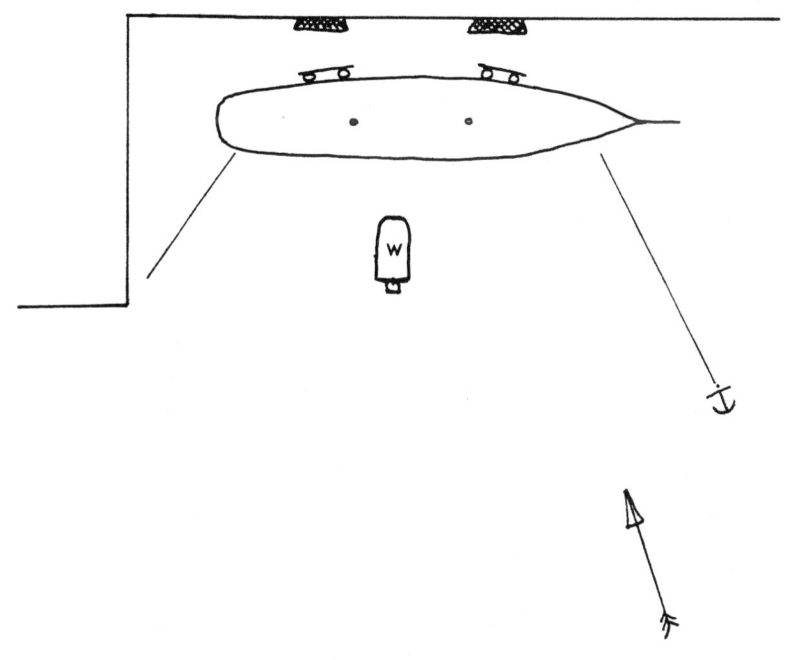

Azara *approaching the dock in a strong onshore wind, going sideways on the wind but controlled by an anchor and a line ashore. Her engine is used to move her fore and aft, so that she lands on the fenders. And the whaler (W) is standing by to nudge her with its bow—where told to—if she is not quite parallel with the dock.*

with all the authorities—customs, immigration, health, and in some cases police—before going ashore. So generally it is best not to dock until you are sure it is allowed. Instead, you anchor where the chart says Quarantine Area, put up a Q flag—a plain yellow one—and wait for instructions.

In Spanish speaking countries, the officials are not well paid and appreciate small gifts—a few packs of cigarettes or a few dollars each—and are mostly interested in checking all your papers, which often takes quite a long time. But in the United States, they are mainly concerned about drugs or other contraband. And if they find any, they can seize the boat. So always be sure that no one is stupid enough to try any amateur smuggling. It is simply not worth it.

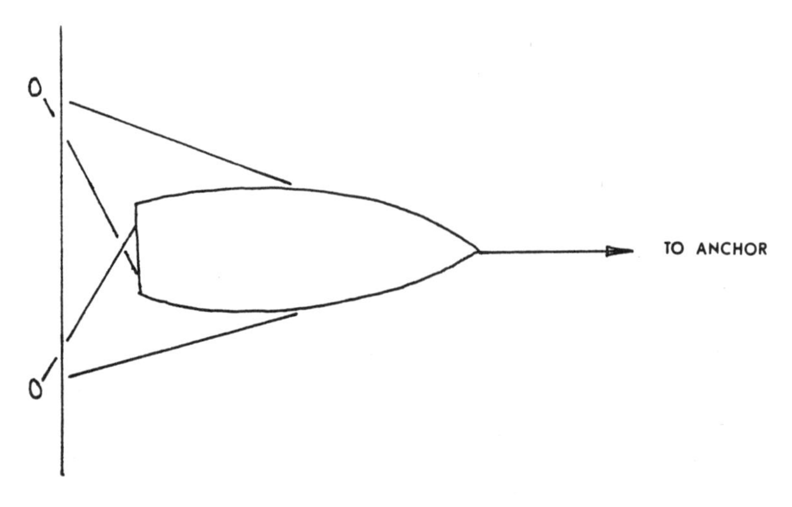

When morning stern to a wall, the stern lines are crossed to make them longer and the spring lines are used to adjust the distance between the boat's stern and the wall.

22

Bound Coastwise

For a coastwise passage from one port to another the first step is to determine what the weather (and therefore the sea condition) will be along your proposed route, then figure your running time and decide when it is best to leave to make the most of the currents and daylight available.

In winter it is wise to be more conservative in planning a passage than you would be in summer, for the weather systems move faster and are stronger. Also there is less daylight and a winter's night at sea can be a cold thing, indeed.

If the sea will be rough, it is good to allow time for a leisurely meal before you start, then some to prepare the boat for the passage in the calm water of the harbor. That is the time to check all fuel filters, to make sure the bilge pump suctions are not clogged, to be certain that everything you might need later will work and then to secure things that could slide, fall, break, or cause damage.

If an anchor must be left on deck (we prefer them

below), it should be lashed down. A small cushion rammed into a locker will keep the things inside it from moving. Radios and such must be wedged in safe, dry places. Make sure the head is empty and its sea cocks are shut. The main valve of a propane tank must be closed, too. Often there are many things to be done.

It is a good idea to work your way down the boat from end to end, looking for anything that could cause trouble while you are at sea and thwarting it before it does.

As we get under way, we stow the docking lines, or anchor rode, leaving nothing loose on deck to fall overboard and get in a propeller. Then we go slowly across the harbor, letting the engine temperature come up, before using more power.

In a motorboat the engines are usually warmed up when we get to the sea buoy, so we bring them up to cruising speed and synchronize them, to even the load. Then we check the gauges and let someone else steer, while we go below and look for trouble in the engine comaprtment. If there is a leak (of oil or fuel or water) or a strange noise or vibration, that is the time to find it—not far out to sea, an hour or more later.

Then the passage itself, in a motorboat, is just a matter of pilotage and supervision, making sure that everything goes as you planned it, that the weather works out essentially as you figured it would, and that each mark comes up on schedule. Usually we check the engines after the first hour and then every two hours, after that. And always we keep an eye lifting to windward, for small (maybe unreported) changes in the weather that might require defensive or evasive action.

With a sailboat, the engine is less important, so it gets less attention. But before leaving the calm water of the harbor, we get ready the sails we will need and if possible we have the mainsail up, to stop her rolling, as we head out to sea.

Then there are many decisions to be made throughout the passage, for although you have figured your course and time to each mark, a sailboat can seldom go there directly. More often you have to tack out to sea, then back inshore, and—depending on how the currents are running—it may pay you to take a few long tacks or many short ones, along the shore.

Crossing a bay from one headland to another, you often find a current along the shore, going in the opposite direction from the main current farther out. So you can carry a fair tide in either direction, except for a short distance—right at the headland—where you may have to go against it.

A slight shift in the direction of the wind may call for a change of tacks, depending on which way you think it will go next. And if the wind speed varies, it may pay you to change a headsail for a larger or smaller one—or even reef the main—to keep the boat moving at her best speed. The wind close inshore is often different, both in speed and direction, from that farther out—especially where there are hills or mountains near the coast—so figuring what route to take, from mark to mark, may not be easy.

Usually it gives you plenty to do. The opportunity to use judgment and skill—plus the quiet and limitless range—is why many people like sailboats. But sometimes you are short handed and just want to get there—as when a husband and wife are moving their boat, before taking a vacation—and in those cases, the art of motorsailing may be useful.

The idea is to make the engine and the wind work

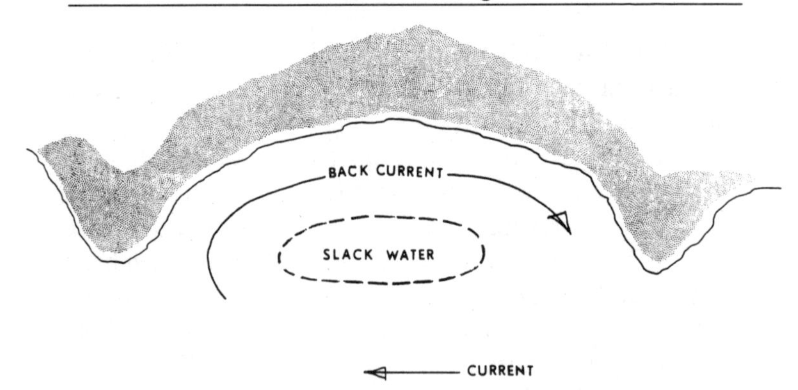

Where a current along a coast goes across a bay—from one headland to another—you often find a back current, close along the shore, which is going in the opposite direction.

together, to produce an effect better than either could alone.

Often all you need do is set the mainsail, sheet it hard in, and get the engine running at its cruising speed. Then the boat will go to windward like a six meter, pointing very close to the wind and moving quickly through the water, with an easy motion—just rising and dipping over the waves.

Only the mainsail can be sheeted flat enough to work that well, going to windward. And because of our forward motion, we seem to be on the wind most of the time, so that is the sail we generally use in boats up to sixty feet long. But in a big boat we sometimes use a staysail and the mizzen, which are easier to hand with a short crew in a sudden squall.

Reaching in a strong wind, we ease the main sheet, to let the sail drive the boat harder, then throttle back the engine a little to save fuel. And running before a strong wind, we may use a Genoa staysail or jib instead of the mainsail. That moves the center of effort forward, which makes the boat handle better and eliminates the boom (also any concern about an accidental gybe). It can easily be handed, if the wind should increase, to be changed for a smaller sail.

So you are still handling a sailboat but now you have one with unusually high performance and a very simple rig.

If you are often short handed, an autopilot might be nice to have but the elaborate ones are usually expensive, while the simpler ones tend to have definite limitations.

The problem is that, before deciding how far to move the rudder, it needs to measure not only the difference in angle between the boat's heading and the desired one, but also which way that difference is changing and how fast. In fact, to be as good as a competent helmsman, it should measure the rate of change of the rate of change and factor that in too.

So the elaborate ones do a good job but you rarely see one aboard a boat less than ninety feet long and the simpler ones, which you find aboard other yachts, are only suitable for use when the steering is easy—in calms or when the seas are abeam or maybe dead ahead. When the steering is hard, as in a quartering sea, they just can not cope with the complex motions.

Loran or Omni can be used on coastwise passages and since both are basically radio receivers, they generally work. Either will give you a position line—if the necessary shore stations are available—but they do it in different ways.

A Loran receiver measures the difference between the time a signal from a shore station arrives and the time another one, from a different shore station, does. But radio signals go very quickly through air and the measurement has to be in millionths of a second, between one event and the other. So taking a measurement can be tricky, unless both of the stations are coming in well. And you may get a spurious signal, especially at night. But if you are fairly sure you have a good one, it is not hard to plot a line of position on a chart with the appropriate curves already printed on it.

An Omni receiver is much simpler, because all the work is done at the transmitting station. It radiates two signals, one omnidirectional and another which varies in phase from that one by an amount equal to the direction of radiation. So if you are northeast of the station, the second signal will be 45 degrees out of phase with the first, if you are south of the station it will be 180 degrees out of phase with it, and so on.

That is a very elegant system. The tuning is digital; you just set the frequency of the station you want, center a needle and read the number of the 'radial' you are on.

But it was made for airplanes and when used at sea, there are problems: You have to transfer the location of the station from an aeronautical chart to a nautical one, before laying off a line of position. And you must remember that the radials are in degrees from magnetic north, at the station. But the biggest drawback is that the frequency used by the system is such that its range is hardly more than forty miles, at sea level. So you do not often find a station available when you want one. And ordinary radio beacons of the kind you take bearings of with a direction finder are usually available at each major turning point, as well as your port of arrival.

23

On the Unexpected

At sea it pays to develop the habit of looking beyond the obvious, for when you stop to wonder why a thing is so, you may find that it gives you some unexpected and useful information on quite another matter of great interest to you.

For example, imagine you are bound for Bermuda out of New York. After a day or so, you put a bucket over the side—to get sea water to boil some potatoes—and you notice that it is much warmer than it was, a few hours ago. Why is that? There is only one answer: you have entered the Gulf Stream—the river of warm water, meandering up from Florida—and that in its turn means you are now being carried to the northeast by the current. So tell the navigator, right away.

Soon the water will be a deep, clear blue with patches of yellow weed on the surface. And the weed may foul the spinner of your log, sooner or later, so listen to the faint noise that the log makes and glance at it now and then to be sure that you will know, within a few minutes, if it stops.

There are other things you might not expect but which are obvious when you stop to think about them. For example, if you are beating to windward in a sailboat and find there is a cross current, it will usually pay you handsomely to get it under the lee bow—by changing tacks, if necessary—for then the course made good will be very much closer to the wind.

Another situation is where you have to turn a long boat in a confined space (as when leaving a dock to go down a river). There it pays to get one end of her out in the middle, where the current is strongest, while the other end is in the slack water, to one side. Then the current will do it for you.

Going down a winding river in a long boat, you may have to start turning the wheel to starboard while the channel ahead of you is still turning to port, to stop her from rotating in time to go straight down the next reach beyond the turn. That is because she has great rotational inertia. But any boat has some, as you can feel by pushing on an end of one that is hanging from a crane over a dock. When you push, she slowly starts to turn. But when you stop, she goes on turning; you have to push (or pull) the other way to stop her.

You get the same effect when steering a boat in a seaway. She may start turning away from her proper heading, so you turn the wheel, until she stops doing it. Then she starts going back but to steady her at the heading you want, you have to give her some rudder the other way, then take it off again.

The important thing to remember is that a rudder does not steer a boat. It merely applies a force that tends to turn her; what she actually does will depend on the sum of that force and all the other ones acting on her at the time. Those include her own rotational inertia, the effects of the seas on her hull, and the wind on her sails. The latter may change as she turns—it probably will—so the situation is a complex one and what to do next is not obvious.

You hear of people having a "feel" for such things and it is true that, with experience, one can hold a good course with little apparent effort. But it comes from thinking about all of the forces that are acting on the boat, being constantly aware of how they are changing, and mentally computing the correction to apply with the rudder to

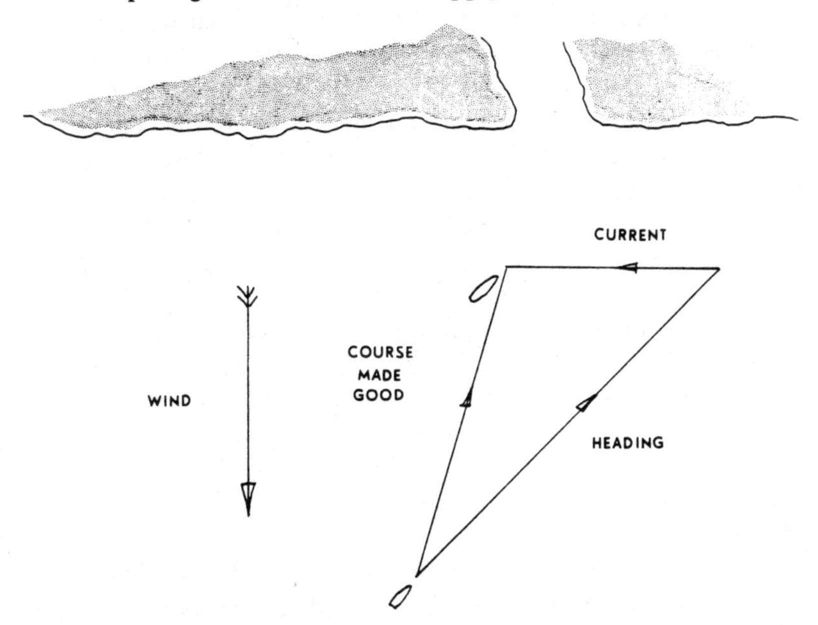

The boat is sailing northeast at 6 knots but the current, under her lee bow, is 3 knots and her course made good is only about 16 degrees from being dead to windward.

balance the equation.

A thing may be unexpected because it comes at a time when it is not supposed to. Everyone is watching out for hurricanes along the east coast of North America from late July until the early part of November. But a hurricane can occur at almost any time of the year and it would be stupid not to pay attention to any word or sign that one might be coming your way.

Hurricanes are carefully watched by the Hurricane Warning Center in Miami (using aircraft and satellites) so the position of one that has already formed—and its probable future track—can be obtained at any time. But in the Caribbean it would be possible to meet one that was just forming which, though less severe than a fully devleoped one, could still be a bad storm.

In that case, the first thing is to find where the center is. Storms rotate anticlockwise—seen from above—but the wind spirals toward the center, by maybe ten degrees, so by measuring the direction of the wind, you can figure the direction of the center of the storm relative to where you are.

Usually it will be moving west or northwest at that time in its life and its forward speed adds to the wind speed on its right side (facing the way it is going) while reducing the wind on its left side. Since the power of the wind varies as the square of its speed, the right side of the storm is clearly the more dangerous one, to be avoided if possible.

A drop in the barometric pressure means that the storm is either intensifying or coming toward you, or both, while a rise (even of a few hundredths of an inch) suggests that it is going away, since it is unlikely to be dissipating yet. So by tapping the barometer to see which way the needle moves, you may get a clue as to whether the storm is coming or going. In your log book, you should find the barometric pressure and wind direction, taken less than four hours ago. And between them, the various clues should enable you to estimate where the storm is centered, as well as where it is going.

With that information you form a plan of action to stay out of harm's way—depending on the type of boat, proximity of land, and so on. Then you watch the barometric pressure and the wind direction to refine or modify your estimate and your plan.

A thing you can expect, now and then, is a micro low. They are small "twisters" like fat, weak tornadoes; the ones that hit us were buzzing around inside ordinary storms. One caught us on the Detroit River in a powerboat and doubled the windshield wipers back into U-shapes but did no other harm as we ran before it at full speed through heavy, driving rain.

Two others appeared when we were about 300 miles southeast of Charleston, South Carolina in a forty-six-foot ketch, lying-to under storm trysail. One hit us, laying her flat on her side, but in a few seconds it was gone and again no harm was done—though the air speed was over 100 miles an hour—because the boat's bottom deflected it over the rest of her, sheltering us.

So it pays to keep an eye lifting, in a storm. If you see a micro low coming, try to avoid it but if you can not do that, tell everyone to hang on tight when it hits you.

Another thing to watch out for is friction among the crew. Most long voyages in boats that fail do so for that reason. The way to avoid it is by careful selection of the people, then setting a tone of polite efficiency—like that in the cockpit of an airliner—in which it is unthinkable for anyone to show any strong emotion, let alone shout or make a fuss.

You can expect the continuing dampness to have odd effects—crossing the Atlantic in *Sopranino,* all our toe-nails softened, then fell off—and for that reason, it would be unwise to trust a "black box" like an electronic calculator offshore, unless you had (and used) something simpler to back it up.

24

Provisioning

When you go aboard a strange boat, you often find a weird selection of things to eat in her lockers. It seems that many people think yachting is a form of camping, so they stock up on convenience foods loaded with chemicals and do not even try to eat as they would at home. Yet on an intracoastal passage or a modest coastwise one it is seldom hard to eat well.

Of course it helps to have a cook aboard. In a tugboat he is paid more than anyone except the captain, which shows what importance the crew attaches to its food. And in extreme cases, we would even recommend marriage to get a good cook.

But even without a cook, it is not hard to put together a decent meal, if you buy sensible things. And the best place to get them is usually a supermarket, where the selection is wide and the prices are reasonable. We have provisioned in them for two people or twenty and for periods of one day to a month, sometimes arriving at the checkout counter with ten baskets full.

Once we asked the manager about a discount for so large a purchase but he told us his profit was only a few cents on the dollar and we believed him, so we tend to think that anyone who offers us a discount may be charging higher prices.

Anyway we shop in supermarkets when we are living ashore, so we have no trouble finding things we like to eat —and have at least *seen* someone cook before. I have a mental block about cookbooks that would stop a river but when you see someone actually cook, you can translate what they are doing into your own terms—like "boil this, while you peel that" and so on.

If we do not know which of several items to buy, we look around for a woman shopper and ask her. Almost always they are very helpful—people like to be asked about their own spheres of expertise—and being customers, they are biased toward the items which they consider best value for the money. You can also ask them for cooking information, like: "How many minutes do you allow per pound of this?"

But if a thing does not seem right when you have cooked it, you can always cook it some more. And the ordinary things, like meat and vegetables and eggs, are unlikely to do you much harm, however badly you may cook them. The point is, they were good, healthy foods to start with and that is basic. Pork must be cooked long enough to be safe (and there is some evidence that beef should be, also), but vegetables usually taste better and do you more good when lightly cooked.

For an inshore passage, when we expect to be at anchor or at a dock each night, we plan meals like those we have at home—except that there may be no oven in the boat and our lunches will probably be eaten under way. At the start, we get what we need for five breakfasts, lunches, and dinners.

Then we look for ice (unless the boat has a refrigerator) and stow the food in the boat, taking care to put dry items in dry places and keep perishable things away from hot places such as a locker adjacent to the engine's exhaust pipe.

For breakfast, we have things like eggs, toast, and coffee. All you need to make toast is a long fork—just spear the bread near an edge and keep it moving, low over the flame. Then lunch is a sandwich or maybe a salad, eaten under way, but dinner—after the boat is secured for the night—is a full meal, with meat and potatoes, a vegetable, and dessert.

Even with the simplest yacht stove (a two-burner one that runs on alcohol) such a dinner is not hard to fix. The trick is to begin cooking each thing at the proper time, so that all of them will be ready at the same moment, then juggle the pots and pans between the two burners and the space between them, where they get some heat from each of the burners.

Having the right pots makes it much easier. A square frying pan with a glass lid is great, because you can often do the vegetable or potatoes in with the meat and keep an eye on them without constantly lifting the lid. Also the lid prevents them from splattering grease all around the stove. Other items we like to have are a saucepan, a coffee pot of the percolating kind, and a deep pot for spaghetti and such. Each one should have a deeply fitting lid to keep the contents inside, if a vessel goes by while you are cooking.

Chicken can be cooked in various ways and near the sea,

you often find good fish. Rice can be served with those, as a change from potatoes. And for dessert you can always cheat by picking something up at the supermarket. So it is not hard to put out an interesting meal, each day of the trip.

But for an offshore passage—where food is important to keep the crew healthy and morale high—selecting it takes more thought and cooking can be difficult, even dangerous.

When departing from the United States, you seldom have a problem getting anything you want but what you can take depends on the boat. The ones capable of offshore passages hardly ever have refrigerators (unless they are very large) so the capacity and insulation of the ice box are limiting factors. A big, well-insulated one may last five days, while a small, poorly insulated one will hardly last three days, especially in hot weather. But in either case, filling it with ice the day before you start will cool it down and packing it tightly with things that are already cold will help it last longer.

Another limitation imposed by the boat is the size of the crew needed to handle her. The more people you have, the sooner the capacity of the ice box will be exhausted. So when shopping at the start, take those factors into account.

The idea is to eat fresh food as long as you can, then go to canned things for the rest of the passage, with vitamins and ascorbic acid as supplements in the later stages.

Fresh meat, milk, and butter must go in the ice box; if there is any space left, leafy vegetables go in next. Hamburger and chicken spoil soonest of the meats. Beef will keep longer, especially if it is frozen hard when you put it in. Butter must be chilled or it melts, and warm milk soon goes bad.

Outside the ice box, leafy vegetables only stay good for a few days but tomatoes last for a week. Carrots, onions, and potatoes last two or three weeks. (The potatoes may have sprouts on them but they do no harm.) Of the fruits, bananas and grapes hardly keep for a week but oranges, apples, and lemons—bought unripe or barely so—often last for two weeks. The best way to store fruit or vegetables is to put them in paper bags (to stop them from moving around) and stow them in the coolest lockers you can find. And eggs, stowed the same way, will keep for two weeks, even in the tropics. Bread will last four or five days in a cool locker and when it is too dried out to eat, you can still toast it (after cutting off any mildewed parts) and sneak it under an egg.

Then, as each thing runs out, you switch to a substitute: canned meats and vegetables; rice (which lasts forever) instead of potatoes; canned fruit for dessert; and so on. For bread, the best substitutes we have found are the large crackers—made of wheat or rye—that come in separate, sealed packages. But sometimes in seaports you can find canned butter that will last the whole trip, without refrigeration.

Be careful about stowing cans beneath the cabin sole, in case the labels come off and clog the bilge pump suctions. But if you do, then mark each one on an end, so that you will know what is in it. Stew made with marmalade—instead of meat—is not much fun to eat. We did that once, at sea.

When provisioning in a foreign country, you may find that very few things you ever heard of—especially in the canned or packaged items—are available. Then you simply have to make do with what there is and adjust your menus accordingly.

Be sure you have enough fuel for the stove and that it is stowed well away from the heat—like in a forward locker. Put three large boxes of matches in waterproof containers, then stow each one in a different part of the boat, so that whatever happens, you will still be able to light the stove.

Do not plan on frying anything at sea (except now and then, in calms) because fried food tends to make people sick and frying at sea can be very dangerous. All you need is one sudden lurch—when the boat is caught by an odd sea—and the cook may have boiling fat thrown from the pan, all over him.

Even boiling water is not a safe operation. We have seen the lid of a kettle go straight up on a column of water that fell back down and went everywhere. But we put on foul weather gear, before cooking at sea—though the temperature below may be over 100—so it ran off, without hurting us. I always cook in a storm, because the skipper handing out hot food is good for the crew's morale, but then I am even more careful than usual, making sure that no cooking pot can move or its lid come off, wiring it on or wedging it with a stick from the overhead, then serving the food in deep cups.

In those conditions, getting the things you want to cook is not easy. When you open a locker, various items come out and start rolling around. But instead of chasing them, it pays you to wait until a thing you want comes by and grab it. One day we saw a sausage fly out of one locker, above the stove, and go into another one across the cabin. So we caught it, coming back.

But in calmer times, it is easier to cook and people are more likely to keep the food down long enough to digest it. So it pays to vary the menus, according to the weather.

Crossing the Atlantic in the twenty-foot sloop *Sopranino,* the motion was wild most of the time, so we ate Basic Stew—canned meat, fresh potatoes, and vegetables, with very little seasoning—which was easy to handle and did not make us sick. But when a momentary calm arrived, we broke out the good stuff. That was 1,400 miles from the nearest land and our dinner was: tomato soup, then partridge with peas and mashed potatoes, then tomato salad, followed by apricots and coffee.

Although bland, simple meals are best in heavy weather, it is important to have "treats" now and then to give the crew hope for the future. Almost anything of a gourmet nature will do—even fancy cookies, if they are sealed in a can.

And after a meal, someone else (not the cook) should clean up the galley—often a revolting job, which encourages people to offer their services as cooks, for the days ahead.

25

Emergencies

The best way to ensure the safety of a boat and her crew is to anticipate every possible emergency, then always have the equipment and procedure ready to meet each one.

So first you consider each kind of emergency and work out the best proceedure for dealing with it (depending on the type of boat, her crew, and so on). Next get the equipment you need to meet the emergency and stow each item where it can be reached quickly. Then make sure that everyone knows exactly what he or she should do—without being told—if any such thing happens.

The most common emergency, in our experience, has been fire at sea. I once saw June put out seven galley fires, while cooking a single meal, about 300 miles from land. The problem was the stove, which ran on alcohol. Since that is soluble in water, she kept a kettle handy to douse the flames. So it was quite harmless, though the galley looked like the gates of hell from where I sat, in the cockpit.

But with a kerosene stove, she could not have used water, because kerosene is not soluble in it and the flames would have spread—floating on the water—into places that might be hard to reach. So she would have used a carbon dioxide extinguisher, if one were available, for two good reasons. The gas does not harm the food—you can continue cooking it—and it is very controllable. A couple of quick shots are all you need, usually, to put out a galley fire. And the rest can be kept for other fires, that may occur in the future.

By comparison, a dry chemical extinguisher has more power to put out fires, for its size and weight. But it makes a mess in a galley and when you use it, a piece of powder may lodge in the valve, causing the propellant gas to escape (over a period of time) and the whole thing will become useless. So we keep them for bigger fires, where it makes sense to use all the contents at one time—as might happen in an engine room.

Another situation where you can not use water is when you have an electrical fire. Then a carbon-dioxide extinguisher is safe to use. So we give it a quick shot with that, then turn off the electricity, and—if we are sure it is off—use water to cool down any wood that may still be smoldering.

That happened aboard the thirty-six-foot yawl *Mehitabel* about 150 miles offshore in a gale at night, so we took the leads off the battery, to be sure the fire would not start again, and went on for several days with no lights. For though a small fire may be relatively harmless, a major one—especially where you can not reach it—could leave you out there with no boat.

When we go aboard a boat, the first thing we do is to note where each fire extinguisher is located (also what kind), then make sure that everyone aboard knows where to find it, how to work it, and just what kinds of fires to use it on.

The automatic systems which you sometimes find in engine rooms of powerboats are fine for looking after things while you are ashore but they generally run on carbon dioxide and if we are aboard, we prefer to control the amount used, instead of letting the whole lot go at one time. So we usually disarm the automatic part and operate the system manually.

When installing a fire extinguisher, it is most important to locate it where you can reach it after a fire has started —for example, by a doorway—and later, we always make sure that each of the extinguishers is properly maintained.

The next most common emergency—in our experience —is a boat trying to sink under you. And usually the first problem is to find out where and how the water is coming in. To find out where, you block the limber holes amidships—in the bilges—with pieces of putty and see on which side of them the water builds up. Then you move them to a fresh location and soon you can tell where the leak is, within a few feet.

Often the water comes through a skin fitting, then leaks from a pipe. So closing the sea cock will stop it. But if the skin fitting is damaged, you may have to block it and for that we like a tapered wooden plug, which can be cut at the proper diameter and driven into the hole. But a piece of broom handle with rags wrapped around it will also do the job.

Another common kind of leak is from the engine's cooling system and that may be hard to find. In the sixty-foot ketch *Orana* we had a bad one, so we went into harbor and put her on a sand bank. Then we found the trouble: a drain cock was open—on the muffler—far up under her after deck. The water, which was under pressure, went quickly through the small hole.

Any boat that has been out of use for a while may produce some interesting leaks but the worst kind are the general ones, which you find in wooden boats that have lain in calm water, so that their planking above the water line has shrunk. Then, as the boat heels, every dried-out seam on the lee side weeps along its whole length, letting in a great deal of water. Such leaks are impossible to fix at sea. My fifty-two foot schooner *Magnet* did that in the North Sea and *Wind Song* tried to sink, in the Sargasso, the same way. But we got *Magnet* into a port and *Wind Song*'s planking swelled up—which reduced and finally stopped the leaking—before we tired of pumping her bilges.

The small electric pumps that you find aboard many yachts are fine for keeping the bilges dry while you are away and can cope with small leaks but anything like that could fail, so you need a manual pump as a backup. And for a serious leak, a pump driven off the engine is good to have. But if necessary you can use the engine itself, as an emergency bilge pump.

To do that, you first close the engine's water intake and disconnect its hose, then attach the bilge pump's suction hose to the engine, so that the water goes through it and overboard, by way of the raw water discharge piping. You need a good strainer on the bilge pump's suction pipe but you should have that anyway (fully ten times its area) and the pipe should be a flexible hose with a chain down to the end of it, so that you can pull it up and clean the strainer.

Chains running through the limber holes—with shock cord near the bow—are handy, because a few pulls on the other end will usually clear all the holes in that row. A piece of canvas that can be hauled down under the hull—to cover a hole while you patch it—is nice to have. And the most common leak of all is at the stuffing box on a propeller shaft. To stop it you need a special wrench, which we always keep near the stuffing box. But when a shaft is turning, there should be a slight leak—just a slow drip—or the gland is too tight.

A less common kind of emergency, in our experience, has been loss of motive power, due to being dismasted in a sailboat or to engine failure, in a boat with no sails available.

We have been dismasted twice in our twenty-foot sloop *Theta*—once in a gale at night in the English Channel and once on Long Island Sound in a thunderstorm. Her rig was experimental, so it was not surprising and each time we continued into port without requiring (or accepting) any outside assistance.

For you can take a tow out of a port and no one thinks it strange but if you take a tow into one, the newspapers say that you were RESCUED, which is embarrassing. And with *Theta,* it was not difficult to set up a very adequate jury rig.

Dismasting usually happens in a storm and the first thing is to be sure that the mast does not sink you by making a hole in the hull. The accepted procedure is to let go everything but the forestay and lie to your wreckage—which is a fine sea anchor—while you get organized. And the next thing we did was have coffee, while we considered the situation.

The mast, boom, sails, and rigging were in a tangled mess, lying just beneath the surface. So we hauled *Theta* up to it and staying on the lee side—so that she would not drift down onto it—we brought her alongside the mast and hauled everything up on deck, then laid it down the middle of the boat.

There was no chance of putting the mast up while the wind and sea were still high, so we secured that and set up the boom as a jury, with the Genoa jib as a mainsail and the working jib (reefed to fit the space available) forward of it. With that rig, she sailed so well that the first time, we went about fifty miles to Fécamp in France and the second time, we beat to windward, up to a dock in Stamford, Connecticut.

Since the heel of *Theta*'s mast was set on deck, it went overboard without doing any damage and each time, we only had to replace a broken fitting, then put the mast back up. Of course, things would not be so easy, in a larger boat but the greater size and weight of her mast would be offset by having a more stable platform to stand on and more winches and other gear to work with (plus a larger crew) so that a similar procedure should usually be feasible, in other cases.

But being dismasted is rare. More often, a loss of motive power happens because of engine failure. Since most powerboats have twin engines, it usually happens when a sailboat is under power with her rig dismantled for an intracoastal passage. Then you have to find the trouble. For that you need a good set of tools—including sockets to free bolts that do not want to turn—and gaskets. Without those, you can not take things off to look inside. Also, we like to have a small meter to measure amps, volts, and ohms in electrical circuits.

When "trouble shooting" it pays to be systematic—taking one thing at a time, making sure it is okay, and going on to the next one—rather than hopping around, trying this or that. Then to fix what is wrong, you have to proceed very carefully to avoid dropping any small parts in the oily bilge water.

But the most useful thing to have, at such a time, is the workshop manual for the engine, with helpful suggestions by the men who built it and pictures of what to do next.

The next kind of emergency, a person falling overboard, has only happened to us once and that was in calm water. In fact she did not fall; she dove in to save a line she dropped overboard. But it was in the early springtime and the water was so cold that she nearly died in the short time it took to bring the boat around in a circle and pick her up.

That is the quickest way to get to a person—slowing as you complete the circle. In a sailboat, it is achieved by gybing, which brings the boat to a stop, close to the point at which the person fell over, if you do it immediately.

The other thing to do without delay is throw the person a life ring, to keep him afloat and make him easier to see. For a head is very hard to find, except in flat, calm water,

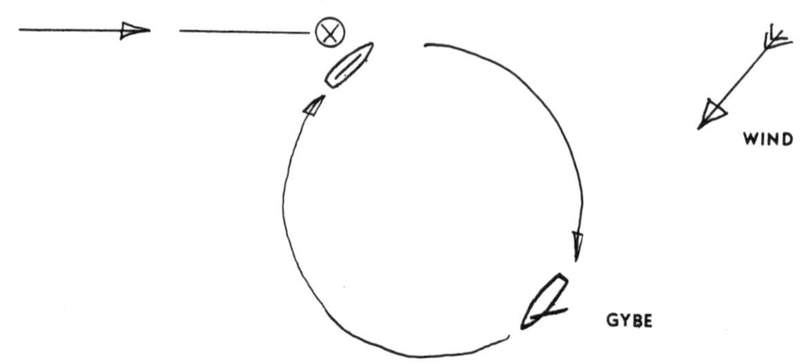

The quickest way to reach someone who has fallen overboard is to come around in a circle, as you slow down. With a sailboat that is achieved by bearing off and gybing her round, which will bring her to a stop—head to wind—alongside him.

even from close range. So the life ring should have a light —attached to it by a lanyard—and preferably a tall buoy, as well. In any case, someone aboard the boat should keep his eyes on the person in the water, as the boat comes round to pick him up, so as not to miss him or run him down, either.

We have picked up several people who fell off other boats—or whose boats sank—and we found it best to come to a dead stop with the person's head where the helmsman could easily see it (amidships or farther aft) then shut down the engines to be sure the propellers are not turning, before throwing a line to him (or more often her) and helping her climb aboard.

But finding a person more than a few yards away, in rough water, is so hard that you might as well fall off a building as off a boat, far out at sea. So we always wear life lines—they need not be fancy, the end of a sheet will do—and make it the responsibility of a man going off watch to see that the one who replaces him puts a line on. Then he will have a good chance of surviving, even if he does slip and fall overboard.

The other kind of emergency is a medical one. Although we have never had any, we have always worried, because they seem to require more training than one can add to a busy life. So we compromised, by learning what we could and always being very careful.

We told our doctor and dentist that we often went to sea and they took care of long term problems, like toothache, on a preventive basis. Then there were some things we could prevent from becoming emergencies, especially burns. It is easy to put your hand on a hot exhaust pipe, when a boat lurches in a seaway. And when you do, the heat removes the fluid from the skin, which is essentially a saline solution. So if you put the hand in a bucket of salt water and keep it there for several minutes, most of the harm is averted.

We had instructions for dealing with a broken bone, until we could get to a doctor, and for appendicitis (keep him warm, feed him nothing but milk, and he should last a week), and we had books on other medical problems, to read if ever the time came, but those were too complex for us to learn by heart.

So if a medical emergency arose at sea, we would certainly try to make contact with a doctor by radio. Unlike a fireman, he can help at a distance. But for all other emergencies, we prefer to be self sufficient, because the kind of thing that causes one very often puts your radio transmitter out of action.

26

Offshore Passages

An offshore passage—where you will be at sea for several days and far from land most of the time—requires a great deal of preparation before you start out. But first of all, you have to decide whether the boat is really suitable for it.

The main things to consider are seaworthiness and range, for she must be able to survive any kind of weather that could arise on the passage, and she must have the range to get where you are going with a reasonable margin to spare.

To be seaworthy, she must be sure to stay afloat—and the right way up—and be able to keep going. So her design must be suitable and her construction and gear adequate. But the strains imposed on hull and gear are functions of the mass and righting moment of the boat, in a given condition. So a wide or heavy one has to be more strongly made than a narrow, light one.

If there is doubt about the boat's ability to survive a

condition that could arise on the passage, a quick phone call to her designer will generally clear it up, besides getting you some useful information on her handling in general.

Sailboats have almost unlimited range, since they can keep going as long as their food and water last. When we crossed the Atlantic in *Sopranino,* we used less than five pounds of food and water—together—per man, each day. So for 28 days, we used no more than 280 pounds of "stores" between two people.

Also, most sailboats are self righting. And since they go slowly, they are unlikely to be holed by anything—like a tree—in the water. So a well-designed and well-built one can often make a safe passage, even if she runs into bad weather.

But a powerboat may not right herself, if heeled beyond a certain angle. And going faster, she is more likely to be holed, if she runs into anything. Her propellers are vulnerable and her upperworks may not stand up to being hit by several tons of fast moving water, as waves break over her in a storm.

So unless she is very large, a typical powerboat can only make offshore passages in calm, stable weather. Before we took the forty-six-foot sport fisherman *Golden Eye* to Bermuda, we waited for three weeks in New Jersey, until a big high pressure dome came across America, then stayed with it all the way. But later, when someone tried to bring her back, she was lost at sea.

And though that passage was only about 750 miles, carrying enough fuel was a problem. Besides full tanks, we needed 1,000 gallons and if we had stowed that in the cockpit, she would have been way down by the stern. So we put it in cans in the cabin and filled her tanks every four hours. Even then, the weight cut her cruising speed from sixteen knots to eleven, so she used more fuel per mile. And if we had carried more, she would have gone slower still, so that was about the limit of her range.

For those reasons, most offshore passages are made in sail and the next step is to consult the pilot charts, which show the winds and currents for each month of the year. Using them, it is not hard to select the best route—which may not be the obvious one—and the best date to start out on the passage.

Sometimes the end of the hurricane season is a factor and while there is no definite answer to that, the insurance people take the first full moon of November as marking that date.

Once you know the date, you can line up a crew. That is the most important thing of all, for more disasters to boats on such passages are caused by crews failing to do what they should, than by all other factors combined.

We never take more people than we need, because extra ones merely get in the way. But the number depends on the boat's rig, as well as her size. A forty-foot ketch only needs three people but in a seventy-foot cutter, it may take seven to reef the mainsail in a sudden squall. So we figure how many we need in the worst case and then we divide them into suitable watches.

Between them, they should have the skills of navigator, engineer, doctor, and cook, as well as those needed to handle the boat. And with a small crew, that means doubling up. Usually the skipper is also the navigator, the mate is the engineer and—if you are lucky—the cook has some medical training.

Clearly it is important that each person has experience in handling similar boats, can steer a good course, and keep watch at the same time. But the ability to get along with other people in a small space is also vital on such a passage. So selecting the best possible crew—and persuading them to take the time to make the trip—is a task worthy of serious effort.

Several days before the boat is due to sail—depending on her condition—the crew should all go aboard to start the long and careful process of checking everything, from stem to stern, from masthead to keel—her hull, gear, machinery, sails, equipment, spare parts, et cetera—then putting aboard the food, fuels (for engine and stove), and water needed for the passage. At the same time, the skipper makes sure that the boat has no outstanding bills—for dockage, chandlery, or whatever—and contacts the various authorities about clearances, so that there will be no unexpected delays at the last moment.

By far the most important item on such a passage is water, because people can only live a few days without it, so we always check the tanks and plumbing for leaks, then disconnect electric pumps which could discharge the whole lot overboard. Next we flush out the tanks, to be sure they are clean and as we fill them, we see what they hold. Then we figure how much we can use each day (with a good margin of safety), and work out a system to keep our total consumption below that. If there are two tanks, we shut one off until later. If there is only one, we put a few cans of water aboard, in case of trouble.

But we do not carry extra fuel for the engine, because the hazard is not worth it, except in special cases—like the long, calm run from Panama to San Diego—and then we put it in rubber containers, made for the purpose and stowed on deck.

When the work of preparation is nearly done, we check

the weather and set a time of departure, then call someone we trust and give him all the details of the passage, as we have planned it—our exact route, time due at each key point, et cetera—so that he can find us, if it should ever be necessary.

At the start of the passage—after you clear the sea buoy and trim your sheets—there is a tendency for everyone to sit around in the cockpit, waiting for something to happen. So it is best to set watches right away and persuade those off duty to try to get some rest, though it may be two or three days, before they adjust completely to living by a new schedule.

The system we like best, for long passages, is the Swedish one, which has five watches a day: From midnight to 4 A.M., then 4 A.M. to 8 A.M., 8 A.M. to 1 P.M., 1 P.M. to 7 P.M., and 7 P.M. to midnight. That way you get plenty of sleep but the early morning watches are short.

And it is good for the skipper to delegate some responsibilities, to take the load off him and to give the other people things of a continuing nature to do, like keeping the equipment in running order and managing the engine's fuel supply.

In the first few days, seasickness is often a problem, caused by fatigue induced by the sudden change of daily schedule and the excesses of the last evening ashore. But rather than use any drugs—which might make one dangerously drowsy—we put up with it. For most of us, it goes away (or reduces to a slight inconvenience) as we settle into our routine at sea.

A more serious problem is sunburn, because that increases, instead of going away. The water reflects the sunlight, so you catch it both ways and often there is no shade available in the cockpit of a boat. But there are ways of dealing with it. From the start—while still preparing to sail—we use creams on our exposed parts and wearing long-sleeved shirts, we slowly build up a tan (through the cloth) that helps protect us. Then at sea, we often wear wide-brimmed hats—tied on with tape—which look unusual but give us shade where we need it most.

An offshore passage often includes some trade wind sailing and if you are reaching, it is fine, brisk going but you have to keep an eye lifting to windward, for tropical squalls. They come down on the wind and rotate anti-clockwise—seen from above—in the northern hemisphere, so the wind you get from one depends on which side of you (north or south) it happens to pass. If it goes north of you, its speed is subtracted from that of the trade wind and you may get a moment of calm. But when one passes close by on your southern side, its speed is added to the wind and for a short while, it can blow very hard. In the southern hemisphere, it is the other way around and of course there are many variations—as when one hits you, fair and square. But once you understand the mechanism, you can tell what to expect and if there is any doubt, it pays to reduce your sail area, in good time, before the squall arrives.

Luckily, they are easy to spot at night, for they blot out the stars, leaving a dark patch of sky. And often you see a line of white on the water—as the wind churns it into spray—which confirms that it is a squall, not just a dark cloud.

But running before a trade wind is tiresome, because there is nothing to stop the boat rolling—we once rolled ninety degrees, out to out, for twenty days—and chafe is a problem. The hanks that hold a headsail to its stay can wear through in a week from the constant motion, so you

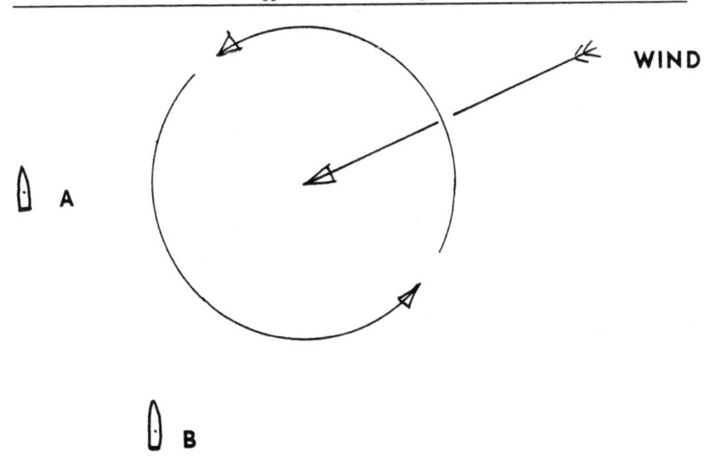

Tropical squalls rotate anticlockwise—seen from above—and move with the trade wind, so boat A will get a wind whose speed is the sum of the two, while boat B may well experience a moment of calm, as the squall passes north of her.

have to take plenty of spares. To sleep, you lie on your back and wedge your head between two pillows, so that it can not flop from side to side. And cooking, or cleaning up the galley, is a less than pleasant occupation.

So if possible, we get the wind slightly off our stern and then the apparent wind comes onto our quarter, which creates the lateral force we need to reduce the rolling. Next, we eliminate the sails with booms and move the center of effort well forward, so that the wind is pulling the boat, not pushing it (in a ketch we use a Genoa jib and a mizzen staysail, nothing else) and that way, we can sail in comfort, with very good control.

On a long passage, it is best if the crew do not know each other too well, so that they can remain strangers and be polite but slightly distant and reserved, because there is no place for emotions in so confined a situation. And if an emergency should arise, the best way to get the crew's attention is to lower your voice and speak in calm, polite, exact phrases.

So life at sea tends to be quiet, with plenty of time for reflection and a thing like washing your socks—by towing them behind the boat, then drying them in the rigging—can become a major item of interest, throughout a whole day.

As you move out of the trade winds, into higher latitudes, you generally go through a narrow band of light breezes and then you come into "normal" weather with variable winds—mostly out of the west—interspersed with calms and storms, as high or low pressure systems and fronts pass over the ocean.

The variable winds produce light to moderate seas and give you the ordinary conditions that you would find, sailing along a coast. But the calms and storms are things that you would not be out in, if you could help it. And there are ways of dealing with each of them but they deserve a piece to themselves.

27

Calms and Storms

Making an offshore passage, it is usual to spend some time becalmed and ride out at least one storm, so it pays to consider in advance how to handle the problems they raise.

As you run into a calm, while there is still enough breeze to steady the boat, the motion is easy, and life is pleasant. You can cook a big meal and eat it in comfort. The jobs that you had put off, because the motion at sea made them difficult, are soon done and before long, everything is squared away.

But when the wind dies, the boat stops moving forward and with nothing to steady her, she rolls on the swell that is left over. If the calm is big, like the one in the Sargasso Sea, you need the sails to get you across it. But if you leave them up, they will slat so hard at the end of each roll that the seams will start coming apart as the stitches break.

So you take them all down and stay on watch, waiting for a tiny breeze. When one comes, you get as many sails up as you can, trim them to it, and hope to gain a few yards, before it goes away. Then you hand them all, to wait again. And that procedure may continue—without a break—for several days.

With the sails down, there is no shade and the temperature in the cabin may be too high for cooking by day. Then we revise our schedule and have dinner at midnight, when it is pleasant to sit in the cockpit, enjoying a leisurely meal.

In a long calm, we worry about water. Will our supply last the passage, with the delay? So we check the figures and—if we must—reduce our daily rations until it is over.

Another problem is chafe. The constant, irregular rolling keeps moving things and where they touch, one may harm another. A halliard can chafe at a sheave; a shroud may cut into a boom. So you freshen the nip at the sheave and glue a piece of rubber to the boom, where the shroud touches it. Then you prowl around the boat, looking for other things to protect.

Usually a sailboat has a limited fuel supply and most of it has been set aside for other purposes—an hour's worth to get you out of the harbor at the start of the passage, an hour a day to charge the batteries, a couple of hours' worth to get you into port at the far end, and some reserved for emergencies. But if there is any to spare, it can be used effectively in a calm. Not all at once but a few hours each day, added to the one for charging batteries, may get you out of it.

And sometimes you are becalmed at the end of a passage, in fifty fathoms or so of water. Then you can use a light anchor, with various spare lines joined together, while the tide is going out of the harbor. And when it turns, you can drift toward the shore, where there should be a breeze in the afternoon time.

But if a storm comes up when we are near a coast with no harbor that we can reach safely, we always head out to sea, for the one thing we need most is room to lie-to.

As the wind increases, it becomes hard to beat into it and running before it is dangerous, because everything seems so calm and peaceful—until you round up into the wind and find that it has become far stronger than you had realized. But reaching is a safe and effective way of getting to a good place.

Even if the wind is on the coast, there is usually one way that a fast reach will take you quickly offshore. But to get on that track, you may have to go about and—if the wind is already strong—you may have to use some special proceedures.

There may be several available—depending on the boat and her rig—and generally we use all that are. So let us consider a large ketch, under all plain sail—jib, staysail, mainsail, and mizzen—to see how the various ones work together.

If the staysail has a boom, we run a line from it down to the lee rail with a slip knot to release it. Then we brief the crew on the procedures we are going to use, so that each one of them knows what to do and just when to do it. Next, we ease the sheets and bear off a little, to get the boat moving fast, as we look for a good spot on the ocean to start the turn.

In a rising storm the waves tend to bunch up into groups, so that you get rough spots—with several close together—and relatively smooth patches of sea between them. So it is best to start your turn at the beginning of one of those. It does not pay to apply too much rudder at first, because that slows the boat down and the main problem, when going about in a storm, is to prevent her from losing headway—to the point where the rudder loses its grip—before her bow crosses the eye of the wind. So you make a smooth turn, increasing the angle of the rudder as your speed through the water decreases.

As she heads into the wind, you glance at the froth in the water beside you to see how much headway you have left. And if she starts going astern—with her bow almost dead into the wind—you change the rudder over but keep it under tight control, so that it can not get away from you and slam hard over.

Then, at the last moment, you haul the mizzen boom across, to kick her stern around. But leave the jib sheet made fast and when the sail goes aback, you have it made. The staysail will go aback, too and her bow will fall off, on the other tack. So you pay out the (now) windward jib sheet, let the staysail's boom go across to the lee side and center the rudder, to help her gather headway, as you trim all the sheets for the new tack.

If she has no mizzen, the main can be used as a kicker, to nudge her stern across. But if you can not get her bow into the wind—to the point where the jib goes aback—you must bear off and start the whole thing again. Then if you can not do it, you may have to wear her around. But we hand the main and mizzen for that, to avoid having to gybe them in such weather.

Once on the proper tack and reaching offshore, you look at the chart for a safe place to lie, while the storm passes over. The main thing is to have room to drift down wind for as long as it may last. But the wind direction will probably change in that time, so first you have to decide what type of weather system is approaching and what part of it will pass over you; then you can figure how and when the wind will change, as it does.

With a sea anchor, a boat drifts almost straight down wind but few people use them any more, because of chafe problems and because nearly all modern boats lie very comfortably under bare poles when the wind is too strong for sailing. And lying-to, a boat will drift as much as forty-five degrees from down wind—in either direction—which gives you a choice of ways to go.

So it is seldom hard to find a place to lie-to, from which you can drift clear of all dangers—including shipping lanes—and when you get there, you heave-to (reducing your sail area as the wind increases) while you get organized, securing things on deck and below, checking everything from lights to bilges to be sure that anything you may need later on will work.

Some people have never tried heaving-to, which is a shame, because it is a most useful maneuver. You can be reaching under a reefed mainsail and a staysail (nothing else) in a rising gale and things are hectic; the motion is jumpy and spray is driving hard across the boat. Then you heave-to, with the staysail aback and everything is so peaceful. The boat goes up and down as the seas roll under her but with no forward motion, she takes little spray aboard. You can stand up and move around with comparative ease and the weather seems to be far less severe.

But the strength of the wind varies with the square of its speed, so when it goes from thirty-five knots to fifty, its power more than doubles and eventually you no longer need any sail to steady the boat and keep her from rolling, so you take it down.

In a sailboat, the Center of Lateral Resistance (CLR) to air is nearly always forward of the CLR to water, so when you take down all her sails, she tends to head off down wind, gathering speed and soon trailing a wake that may

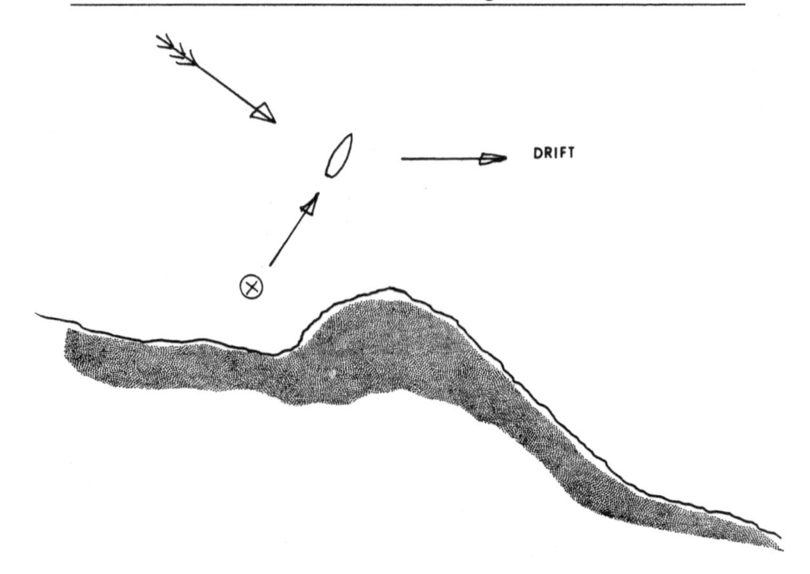

The boat was caught on a lee shore in a rising gale but by reaching fast, she gained some sea room and now she is in fairly good shape, lying-to and drifting away from the coast.

cause a following sea to break over her. But you put the rudder hard over, then back off a bit and secure it, where it can not hit the end stop.

As the boat comes up into the wind, her speed decreases —and with it, the turning effect of the rudder—while the force of the wind on her bow increases until a balance is reached and she settles down, with the wind abeam, making about half a knot in a direction as much as forty-five degrees from down wind.

The storm may last from forty to sixty hours, so she could drift for twenty or thirty miles before it eases. But well offshore, this is not a problem and meanwhile, she will look after herself.

If her bow falls off the wind a little, her speed through the water increases. Then the rudder has more effect and brings her bow back up. So her angle to the wind is essentially stable and the crew can relax, with just one man on watch, to look out for other vessels or anything coming loose on deck.

The wind on the mast stops her from rolling much, so it is relatively comfortable below decks. But if the direction of the wind changes, the boat's course will change with it, so you have to check it, each time you update your dead reckoning. If it is necessary to change tacks, wearing her round under bare poles is very easy: you simply put the rudder over, the other way, and secure it as before, clear of the end stop. The boat will do the rest, coming around and settling on the other tack.

There may well be lightning in a storm and the only thing sticking up, for miles around, is your mast. But protecting the boat from it is not hard: you must provide an easy path for the electricity to follow—down into the water—which takes it as far as possible from the crew. The forestay is a fine route, so if the boat has a metal strip down her stem, all you need do is connect the stay to that with any piece of wire. That has two effects: first, it reduces the chance of the boat being struck; and second, if you are struck, it encourages the lightning to run down the forestay into the sea.

Before you can be struck, a cloud charged with electricity must be overhead. That induces an opposing charge in anything below it, like your boat. But if you drain off the opposing one, you greatly reduce your chance of being struck. And by providing a continuous path from the masthead down to the sea, you let it run off as corona discharge, at about thirty-five microamps, which is too small a current to bother anyone or anything aboard.

Then if a stroke does hit you, a "leader" comes down first which vaporizes your bit of wire but leaves an ionized trail for the main stroke to follow, so that it stays as far away from the crew (and any sensitive equipment) as possible.

If the boat's hull is of steel or aluminum, it will act as a Faraday cage and protect everything inside it. But if it is of wood or plastic and has no metal strip down its bow, we attach a piece of stiff wire, like a coat hanger, to the forestay and run that down her stem, into the sea, as best we can.

During a storm, it is important for people to eat to keep up their strength and only bland food is likely to stay down, so we serve things like soup or stew—in deep, straight-sided cups—three times a day. But alcohol tends to make one sick, as well as reducing one's ability to do what may be needed.

When the wind eases, the rolling of the boat will increase until the mast begins to go past the vertical, as it comes up to windward; that means it is time to go sailing. On deck, the weather still seems pretty bad, with the wind moaning in the rigging and the seas running high. And there is a reluctance to get the first wet sail up. But when it stops flapping and the boat is moving forward, there is a feeling that you will soon be out of the storm and back to normal.

28

Communication

An offshore passage quite often starts from a remote place and when you go there to prepare the boat, you may find you are cut off from many things—like tools, parts, money, and advice—that you are used to having. So it is a good idea to set up your own system for getting them, before you leave home.

The problem is basically one of communication—to get the exact details of what you want to a person who has it, then get it sent to you immediately by the quickest means—and the best "weapons" you have are the telephone and the airplane. But you need someone at home who can always be reached and can understand what you want, who will make local calls until he finds it, then send it off, and see that you get it. That way you can get results that will astonish many people.

For example, someone called us in New York at a quarter to four on a Friday afternoon, to say that he was in Charleston and needed a part for a foreign diesel, which the boat yard could not get in less than ten days. And we had it in his hands—six hundred miles away—exactly two hours and fifty minutes later.

It was not difficult. We phoned the importer in Boston and got the names of his dealers nearby. The third one we called had the part, so we sent a girl to get it. While she was driving, we called various airlines and when she arrived at the dealer, we told her the airport to go to, the airline, and the name of a man who was expecting her. Then she called us with the flight number and time of arrival, the airbill number, and registration number on the airplane. And while it was airborne, we called the man in Charleston, who was there to watch it arrive.

The important thing is that we kept in touch with the part from the time we found it (by phone) at the dealer, until the man who needed it got it. At any moment we knew exactly where it was, for we did not give it a chance to go astray. And the same technique worked again and again, for many years.

But the person handling it needs experience of such things and the determination to follow it through, plus time to spare—at a moment's notice—whenever you call. So finding one may not be easy but it is worth a real effort, especially if you will be preparing a boat for sea in a foreign country.

Even in a Caribbean island, making a telephone call to the United States may mean visiting the phone company's office, then waiting an hour. And often they will only accept collect calls, so the person at home should be forewarned. But when you do make contact, the line is usually clear and you can be sure that what you want has been understood, which makes the phone a far better way to communicate than anything else available.

A telegram, we have found, often goes to the wrong place, or sits in a file for days, before anyone deals with it. But for transferring money, the major credit cards are fine. If you have two of them, you can get a couple of hundred dollars on each, at different banks, in half an hour. For a larger sum it is best to phone home and have the money sent by Telex—to a bank that you specify exactly—then dun them, until you get it.

To get things done in a foreign port, you have to adapt to the outlook of the people. To them, it may seem unreasonable to want a vise to splice a steel wire. Maybe it seldom comes up. Or in Jamaica, for example, you take it to a power station and use the vise there. But no one tells you that—it is too obvious to them—so you must find out by polite questioning.

By the same token, local knowledge or advice may have to be evaluated by your own standards. What is safe? Probably not the same thing to you as to them. So where you know a subject—like how to run your boat—it is best to use your own judgment. But when someone advises you not to jump off a dock (because a shark lives down there), it pays to take his word for it.

Sometimes, in a very remote place, it is better to make do with a substitute than wait for an item to be sent to you. Then local people can be very helpful in suggesting ideas, but in the end you will have to decide what is really feasible.

In countries where most people are poor, theft is usually a problem, so it is not wise to leave a boat unattended, even by day. And in the tropics it often rains hard in the afternoon, but if you leave the hatches closed, the heat in the cabin would be awful. However, if theft is not a problem and you want to leave the boat for a while, some hatches can be made to close themselves—with no one aboard—when it rains.

They must be of the lifting type, as fore hatches commonly are, rather than the sliding kind. But all you need are a small screw eye, an aspirin tablet and a thin rod, for each hatch. Set the screw eye horizontally, inside the coaming, on the side you want open, put an aspirin on top of it and place the rod between the tablet and the hatch, to hold it up. Then, if it rains, the aspirin will melt, the rod will go through the screw eye and the hatch will drop down into the closed position.

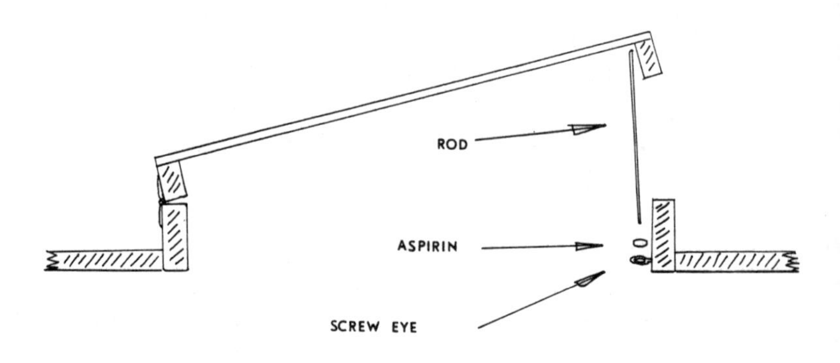

A self-closing hatch: *When rain comes through the opening, it runs down the rod and melts the aspirin tablet. Then the rod goes through the eye and allows the hatch to fall into the closed position.*

29

Using a Sextant

A sextant is a fine, simple instrument, easy to understand and care for, with which you can get a line of position almost any time the sky is clear. All it does is measure the angle from a body in the sky, down to the horizon. But if you take it from ninety degrees, that gives your distance (around the earth) from the body. And since you know where the body was, from an almanac, it is not hard to figure a line of position from it.

To navigate a yacht, you do not need an expensive sextant, because you can not take a sight with such accuracy at sea. But neither do you want a plastic whatsit, made to practice with. We went to a reputable instrument maker and asked for one they had taken in trade, from the mate of a ship, when he was promoted to captain. They had one, already checked out, the price was modest, and we have used it for thirty years with good results.

To let you see the sun (or other body) and the horizon, at the same time, a sextant has a telescope with a split mirror in front of it. On one side, you see the horizon, straight ahead of you. On the other side, you are looking in the mirror and see a second one, in which the sun is reflected. The light goes like a Z from the sun to your eye, when the sun is low.

The second mirror is at the top of the sextant, on the arm that travels around the arc. So by moving the arm, you can vary the angle of the mirror, until the sun and the horizon appear to be at the same height. Then from the position of the arm on the arc—which is marked in degrees—you can see what the angle of the sun was, above the horizon, at that moment.

But a degree makes a difference of sixty miles, so there is a micrometer on the arm for fine adjustment. With that you can read the angle to one minute of arc, which is about one nautical mile, around the circumference of the earth.

In theory, you could read it to a tenth of a minute but if you did, you would just be fooling yourself and adding needless complication to your figuring. If you get within two miles, taking a sight from a yacht on the ocean, you are doing well.

To protect your eyes, there are two sets of filters—one for looking at the sun, the other for the horizon—and each set has filters of different densities, hinged so that you can have as many as you need between you and the light.

For night work there is a light to read by, and to hold it there is a handle (with the battery for the light inside it). On that side are legs to set it down on. But since you are measuring angles to minutes of arc, it is best to keep it in its box—out of harm's way—whenever you possibly can.

In the box, there should be a tommy bar—a small rod with a knob at one end—and on the split mirror are two screws with holes in them. If you put the bar in one and turn it, the object you see in the telescope moves from side to side. And the other screw moves it up or down. So we set the sextant to zero, aim it at a star, and adjust both screws until we see one clear image of it—rather than two, one above or beside the other.

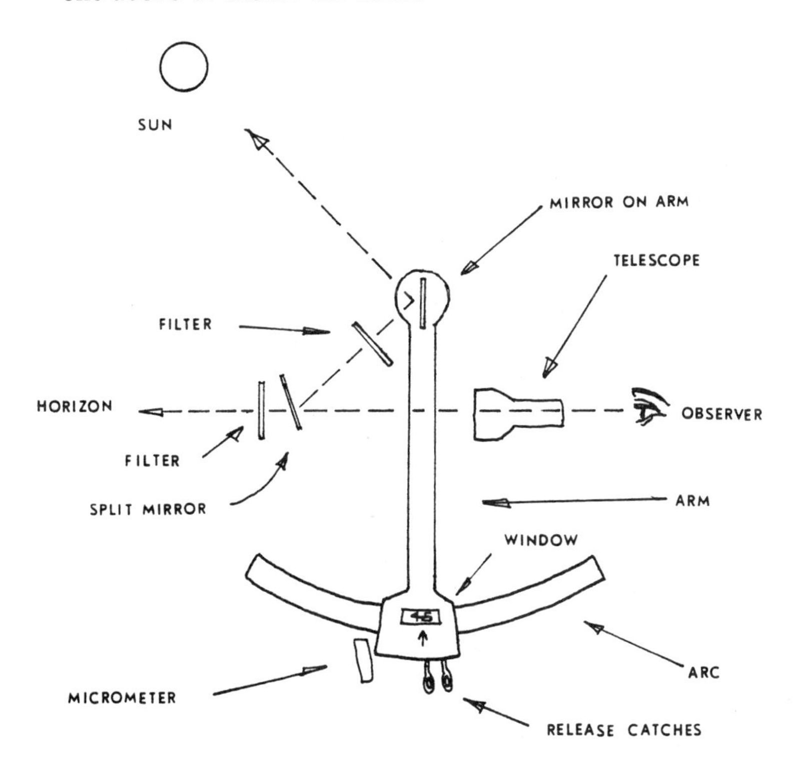

The working parts of a sextant (with frame and handle left out). The observer sees the horizon, straight ahead, through the split mirror and a filter. And he sees the sun, reflected in the top mirror and the split mirror, with a filter between them. The angle, to the nearest whole degree, appears in the window on the arm and the minutes are on the drum of the micrometer.

That corrects any error in the sextant and compensates for our way of looking at things, which eliminates the need for two corrections—one for the instrument, one for the user—when we do our figuring later to find out where we are.

In the box may be another telescope, with a narrower field of view. But on a yacht, the wider one is best, for it is easier to keep an object in it, while she bounces around.

When using a sextant, the main problem is to avoid banging it against anything as you move around the boat. So you have to be constantly aware of it, once it is out of its box.

It should never be lifted by anything except the handle or the frame, so to get it out of the box, you lift the frame with your left hand, then take the handle in your right one. Next you go on deck, to a place where you can see the object whose angle you want to measure and also the horizon, with nothing to get in the way. There you brace yourself, so that you can not slide or fall but the top half of your body—from the waist up—is free to remain vertical, whatever the boat may do.

For a sun sight, it is easiest to set the sextant to zero, adjust the filters, and aim straight at it. With the handle still in your right hand, take the end of the arm (by the catches that release it) in your left one. Then lower the instrument with the right hand and (at the same time) move the arm forward—to keep the sun in view—until the horizon appears in the telescope. At that point you let go the catches, locking the arm.

Now grasp the micrometer, with the thumb and middle finger of the left hand—using the index one, under the arc, to steady the instrument—and adjust it until the sun sits on the horizon like a ball on a table, just barely touching it.

Already you have a rough shot. To improve it, you tilt the sextant from side to side and the sun swings like a pendulum. If it dips below the horizon, you did not have the sextant upright, so you adjust the micrometer and consider the seas.

You only have a valid horizon when the boat is on top of a wave, so you feel her rise and fall, tilt the sextant and as she reaches the top of one—when the deck pauses, before going down—you tilt it the other way. And the sun should barely touch the horizon, as it dips down and swings across it.

If the sun missed the horizon or dipped below it—even by the slightest amount you can discern—that shot was no good, so you adjust the micrometer and try again, from the top of another wave, until you have a good one. Then read the angle and do some more, until you have a series of shots that increase or decrease steadily, as the sun goes up or down. Then you are ready to call out "Stand by," to wake up the man with the watch.

As you take the next shot, you call "Now," and he notes the time (first the seconds, then the rest) and writes it down. Then you read the angle and give it him as two numbers—for example, "Fifteen, thirty six", meaning fifteen degrees, thirty-six minutes—to avoid mistakes. And when you have a few good ones recorded, the job is done. You have all the makings of a position line.

Before putting the sextant in its box, set all the filters in the line of view, or it will not go in. Then look at the list of times and angles, select the shot that is closest to the mean (as they progress), and mark it to be the one you use.

An error of four seconds in the time makes a difference

of a minute of longitude, so you need to rate the watch that you use for navigation and the best way is by checking it against radio station WWV, run by the National Bureau of Standards, which puts out time signals continuously on several frequencies. The first time, you find out the watch's error. Then about a week later, you do it again and the difference between the two errors gives you its rate of gain or loss. And a third check, in another week, tells you if the rate is stable, which it must be, if the watch is to be used for navigation. But the error and the rate are unimportant, as long as you know what they are, so rather than reset the watch, we just keep a log of its error from week to week. Then we can always figure the correct time, to a second, whenever we need it.

In heavy weather taking a sight is harder, because of the motion of the boat. It is not enough to brace yourself; you must cling on, without using your hands. One way is to put a leg over something and tuck the foot behind another, lower down. Or put a foot under something near the deck (like an upturned dinghy) and use your calf muscles to hold the heel on the deck.

You may have to wait for a sea that is not breaking—when it reaches you—for a shot and the readings that you get may be pretty rough. But there is a way of improving on them.

After putting the sextant back in its box, make a graph of all your shots, with the angles on one axis and the times on the other. Then draw a fair line through them and use the shot which is closest to it. Or pick a point on the line and fabricate one, if none of the real ones will do. In any case, the variation you see, between one shot or another and the mean line, will let you know what kind of accuracy to expect in the result.

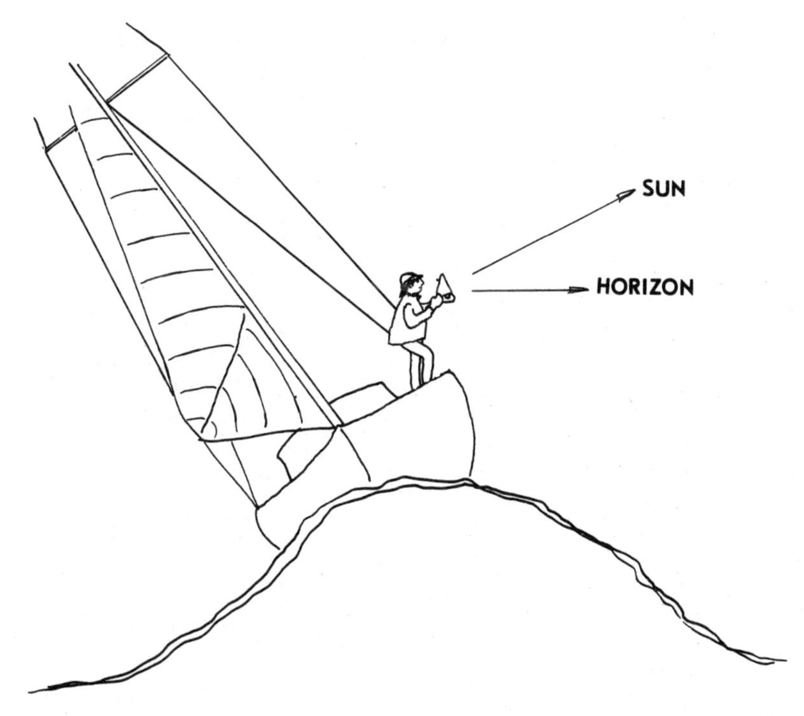

Taking a sight from the top of a wave. One leg is over the life line and tucked behind something, so that the lower part of the body is attached to the boat but the upper part must be free to remain vertical, whatever the boat may do. If the sun were on the other side, you might have to work at the stern.

A moon sight is like a sun sight, except when you do it at night and then the horizon is often hard to be sure of, when you are close to the water. The line between sea and sky seems blurred for maybe five minutes of arc, so we do not trust an angle that we get at night to be accurate within three miles.

The same is true of a star sight but also it is hard to be

sure you are seeing the right one, through the telescope. So you figure your position by dead reckoning, then find what the angle should be, when you take the sight, and set the sextant to that. If you are right, the star will be in the telescope when you aim it at the correct bearing and then you only have to make a small adjustment to the micrometer for your first shot.

That technique is also useful for taking a "snap sight" of the sun, when the sky is overcast but you see it about to appear for a few seconds in a gap in the clouds. With the angle set on the sextant, you wait for the sun to come out and as soon as you can see it clearly, you start taking shots, from the top of each wave. With luck, you may get two or three before it goes behind another cloud, so you graph them to find a mean.

But when the sun is near the edge of a cloud, its rays can be bent by refraction, which would give you a false angle. So be ready to doubt the first shot of a series like that.

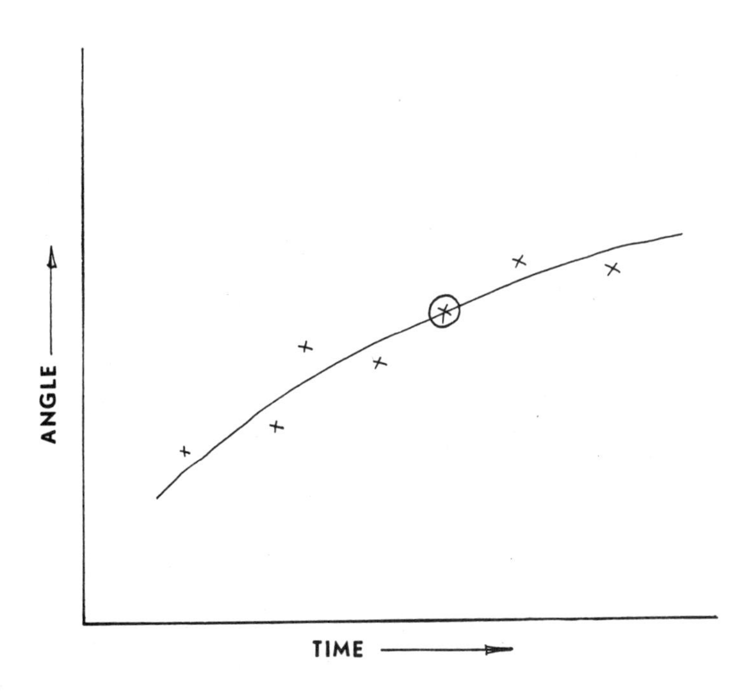

A graph of several shots—like this—makes it easy to see which one to use. And the variation, between one shot and the next, tells you what kind of accuracy to expect.

30

Celestial Navigation

Many people go for months without needing to figure a line of position from a celestial observation, so they tend to forget how to do it. But in some cases, the job is so simple, it can be learned in minutes, once you have the general idea.

When you take the angle of the sun (or other body) from ninety degrees, you get the distance around the earth to the point directly below it, on the surface. If the angle were sixty degrees, it would be thirty degrees away. And so on.

But there are almanacs that say where the sun and moon and stars are, at any given time. The best one to use aboard a yacht is the *Air Almanac,* which is less detailed than the nautical one but gives the data with more precision than you can achieve with a sextant, in average conditions, on the ocean.

Using the almanac, you can figure when the sun will be due south (or north) of your DR position and if you take a sight at that time, all you need do is deduct it from ninety degrees, to find how far away you are, then add (or subtract) the sun's latitude and you have yours. There are some complications but they do not present any problems, when you know about them.

The point on the earth directly below a body is defined by Declination (the same as latitude) and Greenwich Hour Angle (the same as longitude but measured westward from Greenwich) which is called GHA. And the time is Greenwich Mean Time.

So if the watch you use for navigation is on another time, like Eastern Standard, you have to convert that to

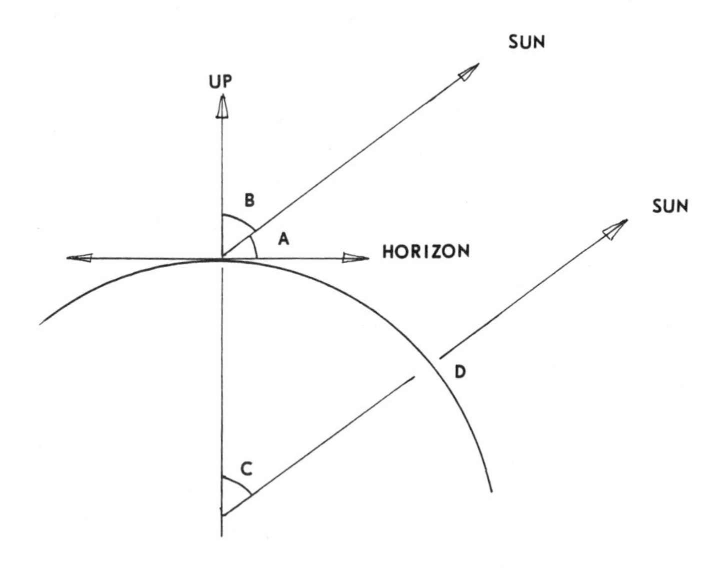

If you take A (the angle of the sun, above your horizon) from 90 degrees, you get B and because the sun is so far away, that is the same as C—the angular distance around the earth, from your position to D. But point D is in the almanac—for any time—so you have the makings of a position line.

GMT—as well as correcting for any error it has at that moment.

We already eliminated two corrections to the angles, taken with the sextant—one for its error, one for our eyes—when we adjusted the instrument. But there are three more.

First we have to allow for the angle between the center of the sun (for which data is computed) and its lower edge, that we had on the horizon. That is its semidiameter and is sixteen minutes, to be added to the reading that we took. Next, we must allow for the "dip" caused by our height of eye (above the sea) which in a yacht is usually between six and twelve feet, for which the correction is three minutes, the other way. And lastly we have a correction for refraction, if the angle is less than ten degrees.

So we avoid taking sights below ten degrees and just add thirteen minutes to the angles that we use—sixteen for semidiameter, less three for dip—which takes care of all the corrections.

In navigator's jargon, the angle is called an Altitude, so for the sun (from a typical yacht) the Observed Altitude plus thirteen minutes is the Corrected Altitude that you will use.

Let us figure a latitude from a noon sight. Imagine we are in the Atlantic and our longitude by DR is 58° 17′ West. We look in the almanac, on the page for that date, and find that at 1540 GMT—3:40 P.M.—the sun's GHA will be 57° 52′. That is given for each 10 minutes of time but inside the front cover, under Interpolation of GHA, we find the sun will take another minute and 39 seconds to go the remaining 25 minutes of longitude. So it will be due south of us at 1541 and 39 seconds, GMT.

We go on deck a few minutes before that (in case our

longitude by DR was off) and take a couple of sights. We do not need anyone to record the time or the readings, in this case, because the sun will go up very slowly, then pause and start to go down. And all we want to know is how high it went. So we keep shooting every now and then, until it starts down, then write the highest good reading that we got on a scrap of paper.

After putting the sextant away, we take the reading, which we will call 78° 22′, and add 13′, which gives a corrected altitude of 78° 35′. Deducting that from 90° leaves us 11° 25′, which is how far north of the sun we are. And adding it to the sun's declination (from the almanac, for that time) which we will call 15° 42′ North, gives us our latitude—in this case 27° 07′ North. This is all so easy that, after doing it a few times on scraps of paper, you can figure it in your head.

To do that, find when the sun will be due south of you (as we just did) and memorize its declination before going on deck. When you get the altitude, add 13′, take it from 90°, add the sun's declination and you have your latitude.

But when you are going back, try to look doubtful, in the hope that someone will say: "All right, where are we?" Then you can tell him the latitude, as you stroll past him.

If the sun is south of the equator and you are north of it (or vice versa), you have to subtract its declination, instead of adding it. But that is obvious, when you think of it. In fact, a virtue of the simple approach to celestial navigation is that if you forget something, you can always find it, by reasoning, from the basic principles involved (which are simple). So there is no danger of losing the things that you have learned.

Another easy way to find your latitude is by using Polaris (the Pole Star) which is so nearly in line with the axis of the earth that its altitude is almost the same as your latitude. The difference (called *Q*) is in a special table in the almanac but to use it, you need the Local Hour Angle of Aries.

Since all the stars go around together, their hour angles are given by reference to an arbitrary point—called Aries and written as a ram's head—whose GHA (longitude) is tabulated in the almanac on the daily pages. It is next to the sun and uses the same interpolations, inside the front cover.

But a Local Hour Angle is the difference, measured to the west, between your longitude and the GHA of anything. So to get the LHA of Aries, you take its GHA and subtract your longitude, when it is west (or add it, if you are east of Greenwich). Then you find Q, for that LHA, in the Polaris Table.

Because of the difficulty of seeing the horizon at night, we take several shots of Polaris and graph them to get a mean. But there is no correction for semidiameter, with a star (they are too far away) so we just subtract three minutes for dip, to get our corrected altitude. Then we add or subtract Q—as it says in the almanac—and we have our latitude.

For a fix, you need another line of position, as well as a latitude, and the easiest way to get it is by taking a sun sight in the morning or afternoon. Clouds often develop after noon, so the morning is a safer bet, except if you are heading east, when your sails may be in the way until nearly midday.

After taking the sight, you add thirteen minutes to the observed altitude, to get the corrected one. Then you express the time of the sight in GMT, corrected for the watch's error.

Now you need Sight Reduction Tables, which are in six books, each covering fifteen degrees of latitude, called HO 229. You can buy them separately and may not need those over sixty degrees. Also, they last for ever, while an almanac only lasts four months.

The tables give the altitude of a body, whose position you know, as seen from different points—about sixty miles apart—so you pick the one nearest to you, then the difference between the tabulated altitude and your observed one tells you how far away it is. So you plot the point, draw a line from it—on a bearing which is also in the tables—measure the distance along it and draw your line of position, at right angles to it.

Each point—which is called an Assumed Position—is at a whole degree of latitude and a whole degree of Local Hour Angle, so you will use the degree of latitude nearest to your position, by DR. To get the LHA, you find the sun's GHA in the almanac and subtract (or add) your longitude in the usual way. But then you round it out to the nearest whole degree.

In the tables, the left hand pages are for when the body's declination has the same name (north or south) as your latitude and the right hand ones are for when they are opposite. Then you look for the latitude and LHA of your assumed position.

There you find columns of figures, headed *Dec, Hc, d,* and *Z. Dec* is the declination of the body (in this case, the sun), *Hc* is its tabulated altitude, *d* is delta (we will come to that later), and *Z* is its Azimuth Angle. Near the top of the page, it says how to convert that to *Zn* (Azimuth) which is the bearing that you will use, when you do the plotting.

The declination is in degrees only but those are not to

be rounded out, so you use the actual number (from the almanac) and look across, for the tabulated altitude, delta, and Z.

Then go to the Interpolation Tables, which are in the ends of the book and look under *Dec Inc* (declination increment) for the minutes of the sun's declination. There you will find, under tens and units, the minutes of altitude to be applied to the tabulated one for each value of delta (plus or minus). When that is done, you have the corrected tabulated altitude.

Now the difference between that and your observed altitude is how far you are, in miles, from your assumed position. So you plot the point, draw a line from it (on the bearing given in the tables), measure off the miles (called the Intercept), and draw in a line of position at right angles to the bearing.

If the angle you observed was less than the tabulated one, you must be farther away from the sun than your assumed position (and vice versa) so you draw the line that way. But it is always safer to keep the whole picture in mind, than to rely on rules that are easy to forget—or to get backwards.

Also, it pays to use common sense in the figuring. If you want to take a longitude of 72° 15′ west from a GHA of 45° 20′, just add 360°—which makes no difference—and you can. If the position line seems to be all wet, go over the figures, looking for a mistake of sixty miles, for the most common error is to lose or gain a while degree, adding or subtracting them.

A morning or afternoon sight, plus a noon one or a Polaris one for latitude, are all that you need on most passages. But if you want to use the moon, stars, and

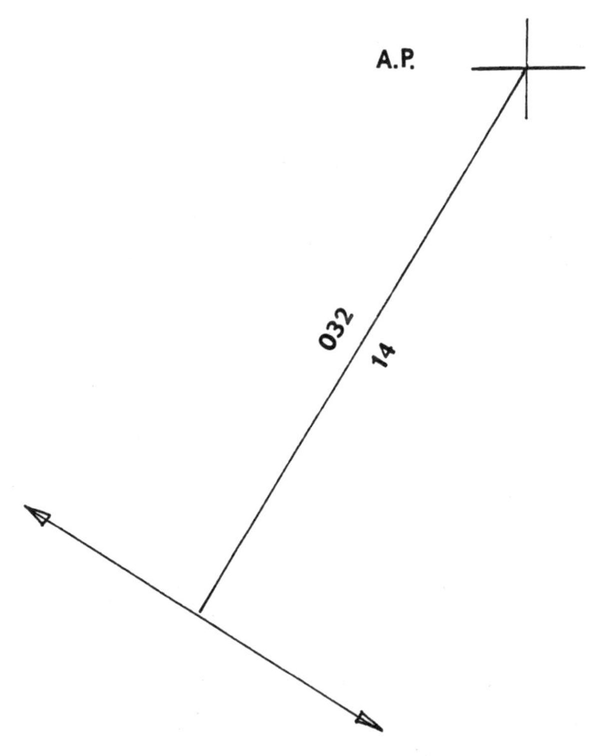

Plotting a line of position: From the Assumed Position, a line (called an intercept) is drawn on a bearing of 032 degrees, for 14 miles—away from the sun, since the observed altitude was less than the tabulated one—then the line of position is at right angles to it.

planets as well, have a chat with someone who navigates small boats across oceans and he will tell you about the corrections in half an hour.

Meanwhile, there is some good stuff (including corrections for other heights of eye and the moon's parallax in altitude) in the *Air Almanac,* which is worth looking through.

And if your watch stops, you can still find your latitude, by starting before noon and taking sun sights until the sun begins to go down. That happened when we were crossing the Atlantic Ocean, so we went down to the latitude of North Point on Barbados a day early and ran in, to raise the light dead ahead.

INDEX

About the Author

Patrick Ellam got his first boat when he could swim across the Thames River in England, fully dressed, at the age of seven. Then for twelve years he spent each summer on the water, graduating from dinghies to cruising yachts. By training he is an engineer and during World War II he taught the workings of antiaircraft fire directors including radar. Then he managed a factory, finding time to sail the North Sea and the Bay of Biscay in a fifty-foot schooner and cross the English Channel nine times in a sailing canoe—once in three hours and five minutes. After starting the Junior Offshore Group—forerunner of the Midget Ocean Racing Club—he sailed the twenty-foot sloop *Sopranino* across the Atlantic Ocean via the Canary Islands and the West Indies to New York, where he worked as chief engineer of a nuclear research company and founded a yacht delivery service employing twenty-five people. Later he set up another firm to import electronic navigational equipment and distribute it across the United States. For years he ran both firms—often sailing on delivery operations—and slowly gained the experience which he now shares with the reader.